Henricus Martellus's World Map at Yale (c. 1491)

Chet Van Duzer

Henricus Martellus's World Map at Yale (c. 1491)

Multispectral Imaging, Sources, and Influence

Springer

Chet Van Duzer
David Rumsey Research Fellow
John Carter Brown Library
Providence, RI, USA

ISBN 978-3-030-08305-2   ISBN 978-3-319-76840-3 (eBook)
https://doi.org/10.1007/978-3-319-76840-3

Softcover re-print of the Hardcover 1st edition 2019

Printed on acid-free paper

This Springer imprint is published by the registered company Springer International Publishing AG part of Springer Nature.
The registered company address is: Gewerbestrasse 11, 6330 Cham, Switzerland

Preface

This study is devoted to a large, late fifteenth-century world map by Henricus Martellus Germanus on permanent display in the Beinecke Rare Book and Manuscript Library at Yale University, New Haven, Connecticut.[1] The map is reported to have belonged to a family in Tuscany for many years before it surfaced in Bern in the late 1950s; it was offered for sale by Franco Novacco, Carla Marzoli, and Van Devanter, and was purchased and anonymously donated to Yale in 1962. It was the subject of preliminary unpublished studies by R. A. Skelton and Roberto Almagià that strongly supported the map's genuineness and importance.[2] The map is thought to have been an important source for Martin Behaim in the creation of his terrestrial globe of 1492,[3] and is also held to be the best cartographic representation of the geographical concepts of Christopher Columbus; in fact, there are good reasons to think that Columbus saw the map or another very similar one by Martellus and that it influenced his ideas about the world's geography.[4] The early reports on the Yale Martellus map called for

[1]The map measures 122 × 201 cm, or about 4 × 6.5 feet, including the printed acanthus border which is 1.2 cm wide and is not an integral part of the map. Its shelfmark in the Beinecke is Art Store 1980.157.

[2]These preliminary studies are now kept in a file about the map in the Beinecke Library (no shelfmark). They are: R. A. Skelton, "World Map by Henricus Martellus Germanus, c. 1489, at Berne," January 10–17, 1960 (18 pages and 1 figure); R. A. Skelton, "World Map by Henricus Martellus Germanus, c. 1489, at Berne: Supplementary Report," June 1960 (5 pages); Roberto Almagià, "Worldmap by Henricus Martellus at Berne," June 1960 (7 pages); and J. S. Sioid, "Documents for the Technical Examination of the World Map Studied in Berne," June 1960 (a collection of natural light, infrared, and ultraviolet images of the map).

[3]See R. A. Skelton, "World Map by Henricus Martellus Germanus, c. 1489, at Berne," January 10–17, 1960 (see note 2), pp. 10–12; George Kish, "Two Fifteenth-Century Maps of 'Zipangu': Notes on the Early Cartography of Japan," *The Yale University Library Gazette* 40.4 (1966), pp. 206–214; Carlos Sanz, "Un mapa del mundo verdaderamente importante en la famosa Universidad de Yale," *Boletín de la Real Sociedad Geográfica* 102 (1966), pp. 7–46, esp. 18–24 and 30; and Arthur Davies, "Behaim, Martellus and Columbus," *Geographical Journal* 143.3 (1977), pp. 451–459.

[4]See Roberto Almagià, "I mappamondi di Enrico Martello e alcuni concetti geografici di Cristoforo Columbo," *La Bibliofilia* 42 (1940), pp. 288–311, esp. 307; particularly R. A. Skelton, "World Map by Henricus Martellus Germanus, c. 1489, at Berne," January 10–17, 1960 (see note 2), pp. 14–17; and Sanz, "Un mapa del mundo verdaderamente importante" (see note 3), 11–18. Davies, "Behaim,

further research,[5] and the map was hailed as being of "supreme importance,"[6] "one of the most precious relics discovered in the twentieth century,"[7] and "a 'missing link' which has long been postulated and sought for by historians dealing with the ideological background to the discovery of America."[8] Yet very little work has been done on the map,[9] particularly not since the burst of interest in the first few years following its discovery and its anonymous donation to Yale.[10]

The main reason why so little scholarly attention has been devoted to the map, which is hand painted, is that most of the texts on it were held to be all but illegible, an observation repeated several times in the literature on the map.[11] While a few of the toponyms in banners and texts in cartouches are readily legible, and some of the longer descriptive texts are in

Martellus and Columbus" (see note 3), went so far as to suggest that the map was made by Columbus's brother Bartholomew, and merely assembled by Martellus; this untenable view is refuted by Ilaria Luzzana Caraci, "Il planisfero di Enrico Martello della Yale University Library e i fratelli Colombo," *Rivista Geografica Italiana* 85 (1978), pp. 132–143, translated into English as "Henricus Martellus' Map in the Yale University Library and the Columbus Brothers," in the author's *The Puzzling Hero: Studies on Christopher Columbus and the Culture of his Age* (Rome: Carocci, 2002), pp. 281–291.

[5] See for example R. A. Skelton, "World Map by Henricus Martellus Germanus, c. 1489, at Berne: Supplementary Report," June 1960 (see note 2), p. 4; Alexander Vietor, "A Pre-Columbian Map of the World, ca. 1489," *Yale University Library Gazette* 37 (1962), pp. 8–12, reprinted in *Imago Mundi* 17 (1963), pp. 95–96, at p. 12/96.

[6] Vietor, "A Pre-Columbian Map" (see note 5), p. 11/96.

[7] Caraci, "Henricus Martellus' Map" (see note 4), p. 282.

[8] See Skelton, "World Map by Henricus Martellus Germanus, c. 1489, at Berne," January 10–17, 1960 (see note 2); quoted by Vietor, "A Pre-Columbian Map" (see note 5), p. 12/96. Skelton closes his Supplementary Report of June 1960, by remarking "Here the Berne map [i.e. the Yale Martellus map] may be compared to the missing piece in a jigsaw puzzle, the identification of which enables other pieces to fall into place."

[9] The lack of scholarly attention devoted to the map is reflected in the brevity of the references to the map in the multivolume *History of Cartography*: see only J. B. Harley and David Woodward eds., *The History of Cartography*, vol. 1: *Cartography in Prehistoric, Ancient, and Medieval Europe and the Mediterranean* (Chicago: University of Chicago Press, 1987), pp. 187 and 316; and vol. 3, *Cartography in the European Renaissance* (Chicago: University of Chicago Press, 2007), pp. 1183 and 1205.

[10] In addition to the works cited in the preceding notes see only Marcel Destombes, *Mappemondes, A.D. 1200–1500* (Amsterdam: N. Israel, 1964), pp. 229–233 with plates 37–38; Sanz, "Un mapa del mundo verdaderamente importante" (see note 3). Some scholars have argued that the "Fourth Peninsula" or "Dragon's Tail Peninsula" in southeastern Asia on Martellus's maps represents a pre-discovery of South America: see Paul Gallez, "Walsperger and his Knowledge of the Patagonian Giants," *Imago Mundi* 33 (1981), pp. 91–93; and Gustavo Vargas Martínez, *América en un mapa de 1489* (Mexico City: Taller Abierto, 1996). This theory has been refuted by Ilaria Luzzana Caraci, "L'opera cartografica di Enrico Martello e la 'prescoperto' dell'America," *Rivista Geografica Italiana* 83 (1976), pp. 335–344; translated into English as "The Cartographic Works of Henricus Martellus and the 'Pre-Discovery' of America," in the author's *The Puzzling Hero: Studies on Christopher Columbus and the Culture of his Age* (Rome: Carocci, 2002), pp. 271–280; and William A. R. Richardson, "South America on Maps before Columbus? Martellus's 'Dragon's Tail' Peninsula," *Imago Mundi* 55.1 (2003), pp. 25–37. The most important recent work on the Yale Martellus map is Bertrand Hirsch, "Connaissance et figures de l'Éthiopie dans la cartographie occidentale du XIVe siècle au XVIe siècle," Thèse de Doctorat, Université de Paris I, 1990.

[11] Skelton, "World Map by Henricus Martellus Germanus, c. 1489, at Berne," January 10–17, 1960 (see note 2), p. 8; Roberto Almagià, "Worldmap by Henricus Martellus at Berne," June 1960 (see note 2), p. 1; Destombes, *Mappemondes* (see note 10), p. 231; Caraci, "Henricus Martellus' Map," p. 282.

fact legible, albeit only through the exercise of great patience. In the vast majority of the texts, the paint used to write the words has faded to a color very close to that of the land or other background. In many spots one can see that there is text on the map, and one can make out the incomplete shapes of a few letters, but there is just too little information to piece together the sequence of letters and read the text. The legends on the map have remained unread despite the late 1950s infrared and ultraviolet photographs, which rendered a few of them legible.[12]

In addition to its connection with the geographical ideas of Columbus and its role as a model for Martin Behaim in the creation of his 1492 globe, there is another reason to study the Yale Martellus map. It has been suggested in a general way that the map (Fig. 1) influenced Martin Waldseemüller in the creation of his famous 1507 world map,[13] the only surviving exemplar of which is held by the Library of Congress (Fig. 2).[14] The two maps use the same projection, a modified form of Ptolemy's second projection with three circular arcs for each meridian, instead of Ptolemy's single circular arc,[15] and the similarity of their depictions of

[12]These photographs, which were made by Professor Cadorin of Basel, may be found in J. S. Sioid, "Documents for the Technical Examination of the World Map Studied in Berne," June 1960 (see note 2).

[13]On Martellus's influence on Waldseemüller, and particularly on the similarities between the Yale Martellus map and Waldseemüller's 1507 map, see for example Sanz, "Un mapa del mundo verdaderamente importante" (see note 3), pp. 17–18; and Peter H. Meurer, "Cartography in the German Lands, 1450–1650," in David Woodward, ed., *The History of Cartography*, vol. 3.2, *Cartography in the European Renaissance* (Chicago and London: University of Chicago Press, 2007), pp. 1172–1245, esp. 1183 and 1205.

[14]On the Library of Congress's acquisition of Waldseemüller's 1507 map see John Hébert, "The Map that Named America: Martin Waldseemüller's 1507 World Map," *Coordinates*, Series B, No. 4 (August, 2003), available at http://purl.oclc.org/coordinates/coordinates.htm. The 1507 map has been published in facsimile in *Die älteste Karte mit dem Namen Amerika aus dem Jahre 1507 und die Carta marina aus dem Jahre 1516 des M. Waldseemüller (Ilacomilus)*, eds. Joseph Fischer and Franz Ritter von Wieser (Innsbruck: Wagner'schen Universitäts-Buchhandlung, 1903; Amsterdam: Theatrum Orbis Terrarum, 1968); and now in John W. Hessler and Chet Van Duzer, *Seeing the World Anew: The Radical Vision of Martin Waldseemüller's 1507 & 1516 World Maps* (Washington, DC: Library of Congress, and Delray Beach, FL: Levenger Press, 2012). There are two excellent high-resolution scans of the map available on the internet site of the Library of Congress. For an account of Waldseemüller and his works see Robert W. Karrow, *Mapmakers of the Sixteenth Century and Their Maps: Bio-Bibliographies of the Cartographers of Abraham Ortelius, 1570* (Chicago: Speculum Orbis Press, 1993), pp. 568–583; and Hans Wolff, *America: Early Maps of the New World* (Munich: Prestel, 1992), pp. 111–126; for discussion of his 1507 map, in addition to the introduction in Fischer and von Wieser's facsimile, see Charles G. Herbermann, "The Waldseemüller Map of 1507," *Historical Records and Studies* 3.2 (1904), pp. 320–342; Elizabeth Harris, "The Waldseemüller World Map: A Typographic Appraisal," *Imago Mundi* 37 (1985), pp. 30–53; and John Hessler, *The Naming of America: Martin Waldseemüller's 1507 World Map and the 'Cosmographiae introductio'* (London: D. Giles, 2008).

[15]Martellus in his map at Yale was evidently the first to use this projection, which permits the parallels to be very close to true scale: see J. B. Harley and D. Woodward, eds., *The History of Cartography* (Chicago: University of Chicago Press, 1987-), vol. 1, pp. 186–187; with corrections in John P. Snyder, *Flattening the Earth: Two Thousand Years of Map Projections* (Chicago: University of Chicago Press, 1993), pp. 33 and 292. Also see Johannes Keuning, "The History of Geographical Map Projections until 1600," *Imago Mundi* 12 (1955), pp. 1–24, esp. pp. 10–11, but Keuning does not mention the Yale Martellus map, as his article was written before its discovery; and Rüdiger Finsterwalder, "The Round Earth on a Flat Surface: World Map Projections before 1550," in Hans Wolff, ed., *America: Early Maps of the New World* (Munich: Prestel, 1992), pp. 161–173, at 165.

Fig. 1 World map made by Henricus Martellus c. 1491 (Beinecke Library, Art Store 1980.157). (Courtesy of the Beinecke Rare Book and Manuscript Library)

Fig. 2 Martin Waldseemüller's world map, Universalis cosmographia secundum Ptholomaei traditionem et Americi Vespucii alioru[m] que lustrationes *(Strasbourg?: s.n., 1507). Library of Congress, Geography and Map Division, G3200 1507 .W3 Vault. (Courtesy of the Library of Congress)*

eastern Asia is particularly striking.[16] Both cartographers extended the lower borders of their maps to accommodate southern Africa, and the maps agree in many other details as well. For example, the depictions of Japan with its many surrounding islands, in both cases at the northeastern edge of the maps, are very similar.

In a talk delivered at the Library of Congress in 2009, I pointed out an additional similarity between the Martellus and Waldseemüller maps, namely that the locations of five cartouches in the western Indian Ocean are identical.[17] This similarity suggested that the contents of these cartouches might be the same, and using the c. 1960 ultraviolet photographs of the Martellus map, I was able to read enough words in two of these legends on the Martellus map to show that they were almost identical to the corresponding legends on Waldseemüller's map. This was tantalizing evidence that the Yale Martellus map, or one very similar to it, was a source used by Waldseemüller in the creation of his 1507 map, not only for its projection and layout but also for its descriptions of places, peoples, and animals.

This likelihood that the Yale Martellus map was used as a source by Waldseemüller in the creation of his important and influential 1507 map, the map that gave the New World the name "America,"[18] cried out for further investigation, for if this use was established, it would connect one of the most important maps of the late fifteenth century with one of the most important of the early sixteenth century. Moreover, a detailed study of Waldseemüller's use of his sources promised to provide insights into his choices as a cartographer and methods of working, insights that would be particularly valuable given the dearth of other information about Waldseemüller and his workshop practices.[19]

---

[16] For discussion of the cartographic history of this peninsula, sometimes referred to as the Cattigara or "Tiger-Leg" peninsula, see George E. Nunn, *The Columbus and Magellan Concepts of South American Geography* (Glenside, PA: Privately printed, 1932), pp. 44–45; and Benjamin Olshin, "*India Meridionalis*: The Dragon's Tail," in his "A Sea Discovered: Pre-Columbian Conceptions and Depictions of the Atlantic Ocean," Ph.D. Dissertation, University of Toronto, 1994, pp. 319–377.

[17] "Evidence for a Lost Map Used by Waldseemüller in his Depiction of Eastern Africa and the Indian Ocean," delivered May 15, 2009, at the conference "Exploring Waldseemüller's World," May 14 and 15, 2009, at the Library of Congress, Washington, DC. A video of the lecture is available at http://www.loc.gov/today/cyberlc/feature_wdesc.php?rec=4569 (it is the third talk in the panel).

[18] See Franz Laubenberger, "The Naming of America," *Sixteenth Century Journal* 13.4 (1982), pp. 91–113; and Christine R. Johnson, "Renaissance German Cosmographers and the Naming of America," *Past & Present* 191 (2006), pp. 3–43. G. C. Hurbult, "The Origin of the Name America," *Bulletin of the American Geographical Society* 18 (1886), pp. 301–316, offers a good discussion of early uses of the name "America" for the New World following Waldseemüller's 1507 map; on this subject also see Fischer and von Wieser, *Die älteste Karte mit dem Namen Amerika* (see note 14), pp. 36–41.

[19] On the Gymnase vosgien or Gymnasium Vosagense, the scholarly association in Saint-Dié of which Waldseemüller was part, see Lucien Gallois, "Améric Vespuce et les géographes de Saint-Dié," *Bulletin de la Société de Géographie de l'Est* 21 (1900), pp. 66–87, especially "Waldseemüller, chanoine de Saint-Dié," pp. 221–229, and "Le Gymnase Vosgien," pp. 88–94. On the Gymnase vosgien also see Gaston Save, "Vautrin Lud et le Gymnase Vosgien," *Bulletin de la Société Philomatique Vosgienne* 15 (1889–1890), pp. 253–298.

In order to pursue this investigation, I obtained new high-resolution natural light, infrared, and ultraviolet images of the map, taken in October of 2010 by John ffrench, Manager of the Department of Digital Media at the Yale University Art Gallery.[20] Of these images, the 2010 natural light images were the most useful. In the natural light images, it is difficult to distinguish the faded text from the similarly colored background, but in a number of cases it is possible, which indicates that some of the text transcribed from the map in this study could theoretically have been transcribed during the last 60 years through patient first-hand study of the map. However, the map's fragility has entailed its housing in a case which renders this type of close study difficult.

In August of 2014, through the support of a grant from the Humanities Collections and Reference Resources Program of the National Endowment for the Humanities, I was able to work with Gregory Heyworth (Lazarus Project, then at the University of Mississippi, now at the University of Rochester), Michael Phelps (Early Manuscripts Electronic Library), Roger Easton (Chester F. Carlson Center for Imaging Science at the Rochester Institute of Technology), and Kenneth Boydston (founder and CEO of MegaVision) to make multispectral images of the Yale Martellus map. Multispectral imaging captures data at specific wavelengths ranging from ultraviolet across the visible bands to infrared, and these data are processed so as to reveal information not visible to the naked eye. This technology, was first developed in order to extract additional information from satellite imagery. In its application to manuscripts and books, it is particularly useful in recovering text that has been rendered illegible due to fading, water damage, fire damage, overpainting, palimpsesting, and cockling (wrinkling).[21]

The fact that the c. 1960 ultraviolet images of the map revealed text that was not visible to the naked eye suggested that the map was an excellent candidate for multispectral imaging, and that turned out to be the case. The multispectral images reveal texts and images on the map that are otherwise invisible, and they reveal many more texts and images than the ultraviolet and infrared images made in 2010. To mention just one example, in the text block in the lower right-hand corner of the map, some words are legible, albeit with the exercise of much patience, in the natural light, infrared, and ultraviolet images of the map, but in the multispectral images made in

[20] I would like to thank John ffrench and also George Miles, Curator of the Yale Collection of Western Americana at the Beinecke Library, for their help in procuring these images. The images made in 2010 have a much higher resolution than the 26 MB color image of the Martellus map available through the Beinecke's library of digital images at http://beinecke.library.yale.edu/digitallibrary/.

[21] For some discussion of the application of multispectral imaging to a manuscript for textual recovery see Reviel Netz and William Noel, *The Archimedes Codex: How a Medieval Prayer Book is Revealing the True Genius of Antiquity's Greatest Scientist* (Philadelphia: Da Capo Press, 2007); for more technical account see William A. Christens-Barry, Roger L. Easton, Jr., and Keith T. Knox, "Imaging and Image-Processing Techniques," in Reviel Netz et al., eds., *The Archimedes Palimpsest* (Cambridge and New York: Published for the Walters Art Museum by Cambridge University Press, 2011), vol. 1, pp. 182–215.

2014, every word of the text is easily legible. It should be mentioned that in many cases where the multispectral images reveal text on the map that is otherwise invisible, reading that text still requires considerable perseverance. In fact, it is often necessary to consult several differently processed multispectral images of the same part of the map in order to read one text.

The multispectral images of the map were made in 55 overlapping "tiles," in 5 rows and 11 columns (see Fig. 3). When multispectral images are referenced in the text below, it will be according to this grid: "C04R03," for example, refers to column 4, row 3. In addition to several multispectral images repro-

Fig. 3 The system of tiles used for the 2014 multispectral imaging of the Yale Martellus map. The tiles are numbered from the upper left corner of the map, which is C01R01, for column 1, row 1. Courtesy MegaVision.

duced in this book proper, Springer has made available electronic supplementary material to the book (see https://link.springer.com/book/10.1007%2F978-3-319-76840-3) that contains multispectral images of all 55 tiles of the map using one of the more successful processing techniques developed by Roger Easton, as well as other multispectral images that reveal specific texts discussed in the book. A convenient way to compare natural light and multispectral images of the Yale Martellus map, and thus to read its heretofore illegible texts, is with the viewer made by Jack Reed, a Geospatial Web Engineer at Stanford University, which is now freely available: see https://www.jack-reed.com/projects/martellus/sidebyside/.

In transcribing Latin texts in this book, abbreviations have been expanded and "u" has been changed to "v," and "i" to "j," where this would help make the word understandable. In transcribing difficult-to-read texts from the Yale Martellus map, the disposition into lines has been retained so as to facilitate comparison of the transcribed text with the multispectral images. Square brackets are used to indicate lacunae, conjectural readings, and editorial comments such as sic.

Eventually, all of the multispectral images of the map made as part of this project, as well as the data captured during imaging at the Beinecke, will be made freely available via the Beinecke's Digital Collections website (http://beinecke.library.yale.edu/tags/digital-collections). The availability of the original data will allow researchers to use newly developed processing techniques to reveal more text and other details of the map without the map itself having to be removed from its case again.

My study of the Yale Martellus map confirms that either this map or one very similar to it made by Martellus was an essential source for Waldseemüller in the creation of his 1507 map, not just for the layout of the map but also for many of the long descriptive texts. At the same time, I show that there are significant differences between the two maps. Some of the differences between them are no doubt the result of Waldseemüller simply choosing not to follow what he found on Martellus's map. But several of the differences suggest convincingly that Waldseemüller was using a different version of Martellus's map. And indeed, it would be too much to expect that the precise map by Martellus that Waldseemüller had in his workshop would have survived.

Although Waldseemüller made heavy use of Martellus for the descriptive texts on his map, he made very little use of Martellus for place names. This is demonstrated through a comparison of the place names on the two maps in the Arabian Peninsula and the western and southern coasts of Africa. Waldseemüller's decision not to use Martellus for most of the place names on the coast of Africa is understandable, as he had a more recent and more detailed record of those names in the world map made in about 1504 by Nicolo de Caverio.[22] His decision not to use Martellus's place names in the Arabian Peninsula is somewhat surprising, as in this case

[22] The Caverio chart is in Paris, Bibliothèque nationale de France, Cartes et plans, SH archives 1. For bibliography on the chart see note 95 in Chap. 1 below.

Waldseemüller did not have a more recent source: On both maps, all of the place names in this region come from Ptolemy's *Geography*.

In addition, this study has revealed that Martellus's depiction of southern Africa is of seminal importance in the history of cartography. The shape of southern Africa on Martellus's map is unusual: It extends far to the east, so that the southern part of the continent has the shape of a shoe, with the toe pointing east (Waldseemüller did not follow Martellus in his depiction of southern Africa, but rather followed Nicolo de Caverio). In the interior, the Nile River system extends much farther to the south than usual, and there are many named cities. Several of these cities also appear on a mid-fifteenth-century map of Ethiopia (which in this context means very approximately Africa south of Egypt) known as the *Egyptus Novelo*, which is preserved as a supplementary "modern" map in three manuscripts of Ptolemy's *Geography*. The *Egyptus Novelo* map is based on geographical data from Ethiopians, probably transmitted to Europe at the Council of Florence in 1441. There is a very important difference, however, between the depiction of southern Africa on the Yale Martellus map and the three surviving *Egyptus Novelo* maps. Those three maps extend several degrees east of the Nile but do not include all of eastern Africa. Martellus copied a version of this map that was more complete than the three *Egyptus Novelo* maps, and that extended this same data further to the east, and so his map is a unique and extremely valuable source in the history of the indigenous cartography of Africa.

Providence, RI, USA — Chet Van Duzer

Acknowledgments

My interest in Henricus Martellus's world map at the Beinecke Rare Book and Manuscript Library grew from an invitation to speak at the conference "Exploring Waldseemüller's World," held in May of 2009 at the Library of Congress—and thus I begin by thanking John Hébert, then Chief of the Geography and Map Division at the library, and John Hessler, then Senior Cartographic Librarian in the same Division, for that invitation. During my initial thoughts and research about the sources Martin Waldseemüller had used in preparing his famous 1507 map, I had a conversation with Monique Pelletier at the Bibliothèque nationale in Paris, and she emphasized that the world map by Henricus Martellus at Yale was an important part of that puzzle that deserved further investigation. I offer her my thanks for that conversation. The similarities of layout between Martellus's map and Waldseemüller's had been noted previously, but no detailed study of the relationship between the two maps had been made, largely because it was almost impossible to read the texts on the Martellus map. I noticed that there are five cartouches in the Indian Ocean off the coast of Africa that are in the same locations on the two maps, which suggested a very close relation between them: since the cartouches were in the same locations, it seemed likely that the texts in them were the same as well.

Through the generous assistance of George Miles, Curator of Western Americana at the Beinecke Library, I was able to obtain scans of ultraviolet images of the Yale Martellus map that had been made in the early 1960s. These images were very exciting, as they revealed large amounts of text on the map in areas where little or no text was visible to the unaided eye. The images enabled me to read a few of the texts on the map, and thus confirm that some of the same texts were in the same locations on Martellus's map as on Waldseemüller's 1507 map. I presented these findings at the 2009 Waldseemüller conference at the Library of Congress, and following the conference Ted Widmer, Director of the

John Carter Brown Library, encouraged me to apply for a fellowship at the JCB to continue my research.

I was fortunate to receive a Jeannette D. Black Memorial Fellowship at the JCB. In support of this study of the map, George Miles arranged for John ffrench, Manager of the Department of Digital Media at the Yale University Art Gallery, to make new natural-light, infrared, and ultraviolet images of the Martellus map in October of 2010. I held my fellowship at the JCB from March to July 2011, and I am grateful to George Miles and John ffrench for making it possible to study the new images of the map, and to the JCB for giving me the time, research facilities, and wonderful collegial ambiance in which to do so. I wish to thank the entire staff of the JCB, particularly Ted Widmer, Susan Danforth, Ken Ward, and Kim Nusco and also the Fellows whose stays at the library overlapped with mine for their camaraderie, particularly José Cárdenas Bunsen for his friendship and support.

Using the images of the map made in 2010, I was able to read a substantial amount of text on the Martellus map, and I used these texts to write a draft of this book about the map. In June of 2012, while I was finishing the book, I had a conversation with Stefanie Walker of the National Endowment for the Humanities, during which I mentioned that I thought that multispectral imaging would reveal substantially more text on the map. She suggested that I apply for a grant under the NEH's Humanities Collections and Reference Resources Program to fund multispectral imaging of the map. I developed a grant proposal with Gregory Heyworth, then of the University of Mississippi, Director of the Lazarus Project, which facilitates the recovery of data from manuscripts by providing researchers with free access to multispectral technology; Roger Easton of the Chester F. Carlson Center for Imaging Science at the Rochester Institute of Technology; Michael Phelps, Director of the Early Manuscripts Electronic Library, which uses digital technologies to make manuscripts accessible for study; and Kenneth Boydston, the founder and CEO of MegaVision, the maker of the camera system, lights, and imaging software used in the project. Bill Christens-Barry of Equipoise Imaging contributed to the early planning for the project, but he was not able to participate in the project itself.

Our application for the NEH grant was successful. Our visit to the Beinecke in August of 2014 to image the Martellus map was coordinated by Raymond Clemens, Curator for Early Books and Manuscripts, and many members of the Beinecke's staff, particularly Christine E. McCarthy, Chief Conservator, and Rebecca Hatcher, Preservation Coordination Librarian, generously made their time available to assist with the project. I thank them enthusiastically for their help. Gregory, Roger, Mike, and Ken are an absolute pleasure to work with, and I thank them for making time for this project—for the preparation, for the imaging session itself, and for the time-consuming processing of the images. Kevin Sacca, a student at the Rochester Institute of Technology, devoted considerable time to the processing of the images and his techniques revealed additional text on the map, and I am grateful to him as well. Nicole Polglaze did additional processing on images of the map during a fellowship at RIT in the summer of 2017, and the results were helpful in clarifying some texts on the map.

I am also grateful to Natalia Lozovsky and Ilya Dines for their suggestions and comments on a draft of the book.

Contents

Preface . . . v

Acknowledgments . . . xv

1 **Henricus Martellus and His Works** . . . 2
Biographical Information . . . 3
Manuscripts of Ptolemy's *Geography* . . . 4
Manuscripts of Martellus's *Insularium Illustratum* . . . 9
Separate World Maps . . . 18

2 **The Legends on the Yale Martellus Map** . . . 44
Legends Outside of the Map Proper . . . 47
Legends Describing the Winds . . . 47
Text in the Lower Left Corner . . . 53
Text in the Lower Right Corner . . . 54
Legends in the Oceans . . . 59
Legends in the Northern Ocean . . . 59
Legend in the Atlantic Ocean . . . 62
Legends in the Western Indian Ocean . . . 63
Legends in the Eastern Indian Ocean . . . 69
Legends in the Proto-Pacific . . . 72
Legends on Land . . . 86
Legends in Northern Asia . . . 86
Legends in Central Asia . . . 104
Legends in Southern Asia . . . 104
Legends in Northern Africa . . . 110

3 **Toponyms in Arabia, Syria, and Mesopotamia** . . . 118

4 **Toponyms on the Western and Southern Coasts of Africa** . . . 124

5 **Southern Africa and the *Egyptus Novelo* Maps** . . . 136

6 **The Influence of the Yale Martellus Map** . . . 166

Conclusions . . . 192

Appendix: Equipment and Techniques Used in the Multispectral Imaging of the Yale Martellus Map . . . 194

Index of Names . . . 202

Index of Tiles of Yale Martellus Map . . . 206

Index of Manuscripts . . . 208

Chapter 1
Henricus Martellus and His Works

BIOGRAPHICAL INFORMATION

Henricus Martellus was a German cartographer active in Florence from about 1459 to 1496. His German descent is clear from his signature on the Yale Martellus map, *Opus Henrici Martelli Germani*, and also in other of his works. In earlier literature it is stated that "Henricus Martellus" is the Latinized form of "Heinrich Hammer," but there is no documentary evidence to show that the cartographer ever used the latter name.[1] Recently Lorenz Böninger as part of his studies of the fifteenth-century German community in Florence has argued that Henricus Martellus Germanus is to be identified as Arrigo di Federico Martello, an employee of the Martelli family of Florence,[2] who were loyal to the Medici and were significant patrons of the arts.[3] However, Luisa Rubini Messerli has disproven this identification with an impressive marshaling of paleographical evidence, and we are left without any knowledge of who Martellus was.[4] The chances are good, though, that Martellus came to Florence from Nuremberg, which was the center of the German Renaissance in the fifteenth and sixteenth centuries, and his work as a cartographer shows the influence of Donnus Nicolaus Germanus, who produced manuscripts of Ptolemy's *Geography* in three different recensions, some of which included *Tabulae Modernae* or new, non-Ptolemaic maps, and who also in 1477 made one terrestrial and one celestial globe for the newly established Vatican Library.[5]

[1] To give just one example, Max Kratochwill, "Henricus Martellus Germanus (Heinrich Hammer)," in Robert Auty et al., eds., *Lexikon des Mittelalters* (Munich and Zurich: Artemis-Verlag, 1977–1998), vol. 4, p. 2138. Other brief encyclopedia articles on Martellus include Franz Wawrik, "Martellus Germanus," in Ingrid Kretschmer, Johannes Dörflinger, and Franz Wawrik, eds., *Lexikon zur Geschichte der Kartographie von den Anfängen bis zum ersten Weltkrieg* (Vienna: F. Deuticke, 1986), vol. 1, pp. 467–468; and Alfredo Pinheiro Marques, "Martellus, Henricus," in Silvio A. Bedini, ed., *The Christopher Columbus Encyclopedia* (New York: Simon & Schuster, 1992), vol. 2, pp. 452–454.

[2] Lorenz Böninger, "Arrigho di Federigho 'Martello': Bürgerknecht, Übersetzer und Kartograph," in his *Die deutsche Einwanderung nach Florenz im Spätmittelalter* (Leiden and Boston: Brill, 2006), pp. 313–348; and Klaus Arnold, "ARIGO - Heinrich Schlüsselfelder aus Nürnberg? Arrigho di Federigho della Magna/Heinricus Martellus in Florenz?" *Pirckheimer-Jahrbuch für Renaissance- und Humanismusforschung* 21 (2006), pp. 161–168.

[3] See Allesandra Civai, *Dipinti e sculture in casa Martelli: storia di una collezione patrizia fiorentina dal Quattrocento all'Ottocento* (Florence: Opus Libri, 1990).

[4] Luisa Rubini Messerli, *Boccaccio deutsch: Die Dekameron-Rezeption in der deutschen Literatur (15.–17. Jahrhundert)* (Amsterdam: Editions Rodopi, 2012), pp. 163–357, esp. 176–214.

[5] On the connection between Nicolaus Germanus and Martellus, see Böninger, "Arrigho di Federigho 'Martello'" (see note 2 in Chap. 1), pp. 334–348. On Nicolaus Germanus's cartographic output see Joseph Fischer, *Entdeckungen der Normannen in Amerika: Unter besonderer Berücksichtigung der kartographischen Darstellungen* (Freiburg: Herder, 1902), pp. 75–90 and 115–121, translated into English as *The Discoveries of the Norsemen in America, with Special Relation to their Early Cartographical Representation*, trans. Basil H. Soulsby (London: H. Stevens, Son & Stiles; St. Louis: B. Herder, 1903), pp. 72–86 and 110–118; Wilhelm Bonacker and Ernst Anliker, "Donnus Nicolaus Germanus, sein Kartennetz, seine Ptolomäus-Rezensionen und -Ausgaben," *Schweizerisches Gutenbergmuseum* 18 (1932), pp. 19–48 and 99–144; Józef Babicz, "Donnus Nicolaus Germanus-Probleme seiner Biographie und sein Platz in der Rezeption der ptolemäischen Geographie," in Cornelis Koeman, ed., *Land- und Seekarten im Mittelalter und in der frühen Neuzeit* (Munich: Kraus

C. Van Duzer, *Henricus Martellus's World Map at Yale (c. 1491)*, Historical Geography and Geosciences,
https://doi.org/10.1007/978-3-319-76840-3_1

Martellus's cartographic works were described some 60 years ago by Roberto Almagià, and more recently by Luisa Rubini Messerli,[6] but a fresh look at his works is required here to explore the interrelationships of his maps and to highlight some features of the works that will prove useful in analyzing his map at Yale.

Martellus's works fall naturally into three groups: his manuscripts of Ptolemy's *Geography*; manuscripts of his *Insularium illustratum*, or island book illustrated with maps; and his separate world maps: a lost world map depicted in a miniature in the Bíblia dos Jerónimos, one of the finest Bibles of the Italian Renaissance; a world map printed by Francesco Rosselli; and Martellus's world map at Yale.

MANUSCRIPTS OF PTOLEMY'S *GEOGRAPHY*

1. Vatican City, Biblioteca Apostolica Vaticana, Vat. lat. 7289. This unsigned and undated manuscript has been dated to just after 1480, as it contains a mention of the Ottoman conquest of Otranto, Italy, in that year.[7] The manuscript is somewhat smaller than average for Ptolemy's *Geography*, measuring 45.7 × 30.5 cm (18 × 12 inches). The maps in the manuscript have been reliably attributed to Martellus,[8] and they are thus the earliest known work by Martellus that survives. The text is the Latin translation of Ptolemy by Iacopo Angeli da Scarperia,[9] and the maps are the standard set of 27 maps of the "A" recension of Ptolemy, without any supplementary *tabulae modernae* or supplementary "modern" maps that

International Publications, 1980) (= *Wolfenbütteler Forschung* 7), pp. 9–42; and Robert W. Karrow, *Mapmakers of the Sixteenth Century and Their Maps: Bio-Bibliographies of the Cartographers of Abraham Ortelius, 1570* (Chicago: Speculum Orbis Press, 1993), pp. 255–265. On the globes made by Nicolaus Germanus see Józef Babicz, "The Celestial and Terrestrial Globes of the Vatican Library, Dating from 1477, and their Maker Donnus Nicolaus Germanus (ca 1420–ca 1490)," *Der Globusfreund* 35–37 (1987), pp. 155–168.

[6] Almagià, "I mappamondi di Enrico Martello" (see note 4 in Front Matter). There is also a brief survey of his cartographic work in Messerli, *Boccaccio deutsch* (see note 4 in Chap. 1), pp. 197–214.

[7] Evelyn Edson, *The World Map, 1300–1492: The Persistence of Tradition and Transformation* (Baltimore: Johns Hopkins University Press, 2007), p. 215.

[8] See Joseph Fischer, *Claudii Ptolemaei Geographiae, Codex Vrbinas Graecvs 82* (Leiden: Brill, and Leipzig: Harrassowitz, 1932), vol. 1.1, pp. 219, 405–408, 489, and 546, and two maps from the manuscript are reproduced in vol. 1.2, plate L35. There is also a brief account of the manuscript in Almagià, "I mappamondi di Enrico Martello" (see note 4 in Front Matter), pp. 290–291, and a brief but good description of the manuscript and a color reproduction of the world map in Marco Buonocore, ed., *Vedere i classici: l'illustrazione libraria dei testi antichi dall'età romana al tardo Medioevo* (Rome: Fratelli Palombi and Rose, 1996), p. 457. Also see Sebastiano Gentile, ed., *Firenze e la scoperta dell'America: umanesimo e geografia nel '400 fiorentino* (Florence: L. S. Olschki, 1992), pp. 242–243 and 247.

[9] On Iacopo Angeli da Scarperia and his translation of Ptolemy's *Geography* see Roberto Weiss, "Jacopo Angeli da Scarperia (c. 1360–1410/11)," in *Medioevo e Rinascimento: Studi in onore di Bruno Nardi* (Florence: Sansoni, 1955), vol. 2, pp. 801–827; reprinted in Weiss's *Medieval and Humanist Greek: Collected Essays* (Padua: Antenore, 1977), pp. 255–277.

incorporate more recent and accurate geographical data.[10] The world map is on folios numbered ff. 1v–2r, and the regional maps are constructed on Nicholas Germanus's trapezoidal projection. Martellus's use of this projection may be ascribed to his association with Nicholas Germanus early in his career, mentioned above; in his later manuscript of Ptolemy, Martellus used a different projection in the regional maps, the traditional Ptolemaic cylindrical projection.

2. Florence, Biblioteca Nazionale Centrale, Magliabechiano XIII 16.[11] This large and beautifully painted manuscript, 57.5 × 42 cm (22.6 × 16.5 inches), was commissioned by Camillo Maria Vitelli of Città di Castello, north of Rome. As Camillo died in 1496, the manuscript was made sometime before that date; on f. 2r there is a miniature inspired by a fresco in Nero's Domus Aurea, which was rediscovered in 1480, so the manuscript was made between those dates.[12] Given the large number of *tabulae modernae* in the manuscript, it is tempting to think that it is from the later part of that date range, as those maps bespeak a considerable development with respect to the Vatican manuscript just described, though the difference between the manuscripts in this regard is also no doubt due to differences in the commissions. On f. 1v there is a small inscription that reads *Henricus Martellus Germanus fecit has tabulas*, indicating that Martellus made the maps in the manuscript. We are to see Martellus's hand in the title on the same folio, as it also refers to the maps[13]:

---

[10] For a brief discussion of the history of adding *Tabulae modernae* to manuscripts and printed editions of Ptolemy, see James R. Akerman "From Books with Maps to Books as Maps: The Editor in the Creation of the Atlas Idea," in Joan Winearls, ed., *Editing Early and Historical Atlases* (Toronto: University of Toronto Press, 1995), pp. 3–48, esp. 8–13; for a more detailed discussion see Akerman's "On the Shoulders of a Titan: Viewing the World of the Past in Atlas Structure," Ph.D. Dissertation, Pennsylvania State University, 1991, pp. 228–253.

[11] For discussion of the Magliabechiano manuscript, see Enrico Narducci, "Opere geografiche esistenti nelle principali biblioteche governative dell'Italia," in *Studj bibliografici e biografici sulla storia della geografia in Italia* (Rome: Tip. Elzeviriana, 1875), pp. 391–470, esp. 404–406; Joseph Fischer, *Claudii Ptolemaei Geographiae* (see note 8 in Chap. 1), vol. 1.1, pp. 219 and 398–404, and two maps from the manuscript are reproduced in vol. 1.2, plate L36; Almagià, "I mappamondi di Enrico Martello" (see note 4 in Front Matter), pp. 291; Sebastiano Gentile, *Firenze e la scoperta dell'America* (see note 8 in Chap. 1), pp. 240–243 with plates 47–48, including a good list of the *tabulae modernae* and bibliography; Guglielmo Cavallo, ed., *Cristoforo Colombo e l'apertura degli spazi: mostra storico-cartografica* (Rome: Istituto Poligrafico e Zecca dello Stato, Libreria dello Stato, 1992), vol. 1, pp. 517–521, with a good color reproduction of the world map on pp. 518–519; and Lorenz Böninger, "Zur Ptolemäus-Renaissance bei Henricus Martellus," in Werner Kreuer, ed., *Monumenta cartographica: Tabulae modernae: kartographische Denkmäler, ein Triumph über die Zeit: Essener Bearbeitung von sechs Tafeln der historischen Kartographie mit sechs vollfaksimilierungen aus: Biblioteca Nazionale Centrale, Florenz (Tafeln 1, 2, 3, 4, 5), Bayerische Staatsbibliothek, München (Tafel 6)* (Ludwigsburg: Edition Libri Illustri, 2001–2003), pp. 11–15. The manuscript has been published in facsimile as *Ptolomei cosmographia* (Florence: Vallecchi, 2004), with studies by Sebastiano Gentile and Angelo Cattaneo. The "Introduzione" by Cattaneo, pp. 23–53, is a valuable description and analysis of the manuscript and its maps.

[12] For discussion see Cattaneo's "Introduzione" (see note 11 in Chap. 1).

[13] This title is transcribed in Fischer, *Claudii Ptolemaei Geographiae* (see note 8 in Chap. 1), vol. 1.1, p. 398; Gentile, *Firenze e la scoperta dell'America* (see note 8 in Chap. 1), p. 241; and by Cattaneo, with an Italian translation, in his "Introduzione" (see note 11 in Chap. 1).

Cl. Ptolomei Cosmographia cum tabulis regionum n[ost]ri t[empor]is et universis portubus et locis maritimi tractus tam notis quam a rege Portus Galli nuper repertis: hoc ornatissimo codice continentur.

In this very splendid manuscript are contained the Cosmography of Claudius Ptolemy with modern regional maps and all the ports and coasts, both those already known and those newly discovered by the King of Portugal.

This title clearly indicates Martellus's interest in the new discoveries of the Portuguese, an interest of which we will see evidence in other of his works discussed below. In addition to the traditional 27 Ptolemaic maps (including the world map on ff. 88v–89r) (Fig. 1.1), the manuscript has 12 *tabulae modernae* which depict the British Isles (f. 91v), the Iberian Peninsula (ff. 94v–95r), France (ff. 98v–99r), Central Europe (ff. 102v–103r),[14] Scandinavia (ff. 104v–105r), Italy (ff. 110v–111r), Sardinia, Sicily, Corsica and Cyprus (f. 113v), the Balkan Peninsula (ff. 120v–121r), Crete (ff. 122v–123r), Asia Minor (ff. 134v–135r), the Holy Land (ff. 142v–143r), and northwestern Africa (ff. 158v–159r).[15]

This is a remarkable number of *tabulae modernae*, a far greater number than appeared in the manuscripts of Ptolemy with maps by Donnus Nicolaus Germanus[16] or Pietro del Massaio,[17] and these maps testify to Martellus's strong interest in acquiring new and more accurate cartographic knowledge. Some of the *tabulae modernae* are very similar to maps in manuscripts of Martellus's *Insularium illustratum*, and indeed some of the texts that accompany the *tabulae modernae* in the Florence manuscript of Ptolemy are the same as the texts that accompany the corresponding maps in manuscripts of the *Insularium*.[18] For example, on f. 88r just preceding the world map, there is a text titled "Descriptio orbis Pii secundi

[14]The map of Germany or Central Europe, which also appears in the Florence manuscript of Martellus's *Insularium*, is thought to derive from a map by Nicholas of Cusa, on which see Joseph Fischer, *Die Karte des Nicolaus von Cusa (vor 1490) die älteste Karte von Mitteleuropa* (Prague: Kommissionsverlag K. André-Staatsdruckerei, 1930) = *Kartographische Denkmaeler der Sudetenländer* 1 (4 pp.); Joseph Fischer, *Die zur Cusanus-Karte gehörige Descriptio Germaniae Modernae* (Prague: Geographisches Institut der Deutschen Universität, 1936) = *Kartographische Denkmaeler der Sudetenländer* 10 (7 pp.); Dana Bennett Durand, *The Vienna-Klosterneuburg Map Corpus of the Fifteenth Century: A Study in the Transition from Medieval to Modern Science* (Leiden: E. J. Brill, 1952), pp. 256–259 and plate 22; and Peter Meurer, *Corpus der älteren Germania-Karten: ein annotierter Katalog der gedruckten Gesamtkarten des deutschen Raumes von den Anfängen bis um 1650* (Alphen aan den Rijn: Canaletto Uitgeverij, Repro-Holland, 2001), pp. 78–84.

[15]There is some discussion of the *tabulae modernae* in the Magliabechiano manuscript in Johannes Werner Kreuer, "Die kartographische Wende von Schedel zu Martellus: Descriptio und Karte," in Dagmar Unverhau, ed., *Geschichtsdeutung auf alten Karten: Archäologie und Geschichte* (Wiesbaden: Harrassowitz, 2003), pp. 129–146; also see Cattaneo's "Introduzione" (see note 11 in Chap. 1).

[16]On the manuscripts of Ptolemy with maps by Donnus Nicolaus Germanus, see above, note 5 in Chap. 1.

[17]On the manuscripts of Ptolemy with maps by Pietro del Massaio, see Germaine Aujac, "Le peintre florentin Piero del Massaio et la 'Cosmographia' de Ptolémée," *Geographia Antiqua* 3–4 (1994–1995), pp. 187–204; and Louis Duval Arnould, "Les manuscrits de la *Géographie* de Ptolémée issus de l'atelier de Piero del Massaio (Florence, 1469 - vers 1478)," in Didier Marcotte, ed., *Humanisme et culture géographique à l'époque du Concile de Constance autour de Guillaume Fillastre: Actes du Colloque de l'Université de Reims, 18–19 novembre 1999* (Turnhout: Brepols, 2002), pp. 227–244.

[18]See Gentile, *Firenze e la scoperta dell'America* (see note 8 in Chap. 1), p. 243.

Fig. 1.1 The world map in the manuscript of Ptolemy's Geography *made by Martellus and now in Florence (Biblioteca Nazionale Centrale, Magliabechiano XIII 16, ff. 88v–89r). By concession of the Ministero dei Beni e delle Attività Culturali e del Turismo/Biblioteca Nazionale Centrale, Firenze. (Further reproduction in any medium is forbidden)*

pontificis maximi" which is an excerpt from Chapter 2 of the *Historia rerum ubique gestarum* of Enea Silvio Piccolomini (later Pope Pius II), which was first printed in 1477.[19] This text quoted from Piccolomini addresses the question of whether the *orbis terrarum* (the inhabited part of the world) is circumnavigable,[20] and this same text appears with the world maps in some manuscripts of Martellus's *Insularium* (see below). Thus we see that Martellus used two different genres, Ptolemy's *Geography* and Cristoforo Buondelmonti's *isolario*, as foundations on which to build something similar to a modern atlas.

On the Yale Martellus map, which will be discussed in detail below, the toponyms in the Arabian Peninsula come from Ptolemy, and if we compare them with those on the map of Arabia in the Magliabechiano manuscript of Ptolemy (ff. 146v–147r), we find an interesting difference: the spellings of some of the toponyms are different and in a consistent way. Specifically, the spellings in the Magliabechiano manuscript are more faithful to the Greek originals, as they have more "h"s reflecting the original thetas and chis. Thus in Arabia, moving from north to south, we have *obraca* on the Yale Martellus map versus *obracha* in the Magliabechiano manuscript, *alata* versus *alatha*, *chargata* versus *chargatha*, and *sabata* versus *sabatha*, among other examples. Thus Martellus was using different versions of Ptolemy for the Ptolemaic toponyms on his map at Yale and for the Magliabechiano manuscript. I have not found a printed edition of Ptolemy's *Geography* in which the spellings of the toponyms match those in either the Magliabechiano manuscript or the Yale Martellus map, so Martellus used one manuscript of Ptolemy as a source for the Yale map and another as a source for the Magliabechiano manuscript. Given that only about 1 out of 16 manuscripts from the fifteenth century survives,[21] it is likely that neither of the two manuscripts that Martellus used survives, but it would certainly be interesting to know why he lost confidence in one of them, or whether it was simply the case that he had convenient access to one while making the Magliabechiano manuscript and to another while making his map at Yale.

[19] For discussion of this work of Piccolini's, see Nicola Casella, "Pio II tra geografia e storia: La 'Cosmografia,'" *Archivio della Società Romana di Storia Patria* 95 (1972), pp. 35–112.

[20] For discussion of this text in the Magliabechiano manuscript, see Gentile, *Firenze e la scoperta dell'America* (see note 8 in Chap. 1), p. 243, and Cattaneo's "Introduzione" (see note 11 in Chap. 1).

[21] Eltjo Buringh, "Losses of Medieval Manuscripts," in his *Medieval Manuscript Production in the Latin West: Explorations with a Global Database* (Leiden and Boston: Brill, 2011), pp. 179–251, esp. 231–232.

MANUSCRIPTS OF MARTELLUS'S *INSULARIUM ILLUSTRATUM*

There are classical precedents for the medieval and Renaissance genre of the island book or *isolario*,[22] but the earliest surviving medieval example was written between 1385 and 1410 by Domenico Silvestri, a friend of Boccaccio's, and was titled *De insulis et earum proprietatibus* (On Islands and their Properties).[23] The first island book to be illustrated by maps was that of Cristoforo Buondelmonti, the *Liber insularum Archipelagi* (Book of the Islands of the Archipelago), of which the author created four different versions, ca. 1418, 1420, 1422, and ca. 1430. The book addresses about 79 Aegean and Ionian islands, and also the peninsulas of Mount Athos and of Constantinople and Gallipoli, and exists today in about 70 manuscripts.[24] Bartolomeo dalli Sonetti created an *isolario* that exists in three undated manuscripts and was printed in Venice c. 1485[25]; the next *isolario* created following Sonetti's was that of Martellus, which is based on Buondelmonti's and survives in six manuscripts.

Martellus added some Mediterranean islands to Buondelmonti's selection, including Corsica, Sardinia, Sicily, and Cyprus, and his *isolario* is the first to include islands outside the Mediterranean (Britain, Ireland, Taprobana, and in one case, Japan[26]), regional maps (of the eastern Mediterranean, Black Sea, Western Europe), and is also the first of which a

[22] On the classical precedents of the *isolario*, see Paola Ceccarelli, "I *Nesiotika*", *Annali della Scuola Normale Superiore di Pisa, classe di lettere e filosofia* 19.3 (1989), pp. 903–935; for general discussion of the genre, see George Tolias, "*Isolarii*, Fifteenth to Seventeenth Century," in *The History of Cartography*, Volume 3, *Cartography in the European Renaissance*, ed. David Woodward (Chicago: University of Chicago Press, 2007), part 1, pp. 263–284.

[23] Domenico Silvestri, *De insulis et earum proprietatibus*, ed. C. Pecoraro = *Atti della Accademia di scienze, lettere e arti di Palermo* 14.2 (1954), pp. 1–319; Marica Milanesi, "Il *De Insulis et earum proprietatibus* di Domenico Silvestri (1385–1406)," *Geographia Antiqua* 2 (1993), pp. 133–146; José Manuel Montesdeoca, *Los islarios de la época del humanismo: el 'De Insulis' de Domenico Silvestri, edición y traducción* (La Laguna: Servicio de Publicaciones, Universidad de La Laguna, 2004) (CD-ROM edition).

[24] For discussion of Buondelmonti's *Liber*, see Hilary Turner, "Christopher Buondelmonti and the Rise of the Isolario," *Terrae Incognitae* 19 (1988), pp. 11–28; and Giuseppe Ragone, "Il *Liber insularum Arcipelagi* di Cristoforo dei Buondelmonti: filologia del testo, filologia dell'immagine," in Didier Marcotte, ed., *Humanisme et culture géographique à l'époque du Concile de Constance. Autour de Guillaume Fillastre. Actes du Colloque de l'Université de Reims, 18–19 novembre 1999* (Turnhout: Brepols, 2002), pp. 177–217.

[25] For discussion of Sonetti's *isolario*, see Wouter Bracke, "Une note sur l'*Isolario* de Bartolomeo da li Sonetti dans le manuscrit de Bruxelles, BR, CP, 17874 (7379)," *Imago Mundi* 53 (2001), pp. 125–129; and Massimo Donattini, "Bartolomeo da li Sonetti, il suo *Isolario* e un viaggio di Giovanni Bembo (1525–1530)," *Geographia Antiqua* 3–4 (1994–1995), pp. 211–236. There are two facsimile editions of Sonetti's book: Bartolommeo dalli Sonetti, *Isolario* (Amsterdam: Theatrum Orbis Terrarum Ltd., 1972); and Bartolomeo dalli Sonetti, *Isolario* (Valencia: Vicent García, 2006), the latter of a hand-colored copy of the work.

[26] For discussion of Martellus's map of Japan in the Florence manuscript of his *isolario*, see Almagià, "I mappamondi" (see note 4 in Front Matter), pp. 298–299, with a reproduction on p. 303; and George Kish, "Two Fifteenth-Century Maps of 'Zipangu': Notes on the Early Cartography of Japan," *The Yale University Library Gazette* 40.4 (1966), pp. 206–214.

world map is an integral part.[27] Thus Martellus's *isolario* represents a bold and important step toward the development of the atlas.[28] Martellus shows himself very interested in obtaining the latest and most correct cartographic data.[29] For example, in the middle of Scotland, he has written *Haec est uera proportio istius insulae*, "This is the true proportion of this island," implying a rejection of an earlier cartographic image of the island and the generation of a new, revised image.[30] We will see other evidence of Martellus's concern with accuracy in the descriptions of the manuscripts of his *Insularium* that follow.

The six surviving manuscripts of Martellus's *Insularium* are undated, but three of the five may be placed in chronological order according to their texts and the development of their world maps, and we may tentatively place the manuscripts in chronological order as follows:

1. Florence, Biblioteca Medicea Laurenziana, Pluteo 29.25. The manuscript has 97 maps numbered in Arabic numerals, followed by 14 maps numbered in Roman numerals. The text begins *Christophori Ensenii ad Reverendissimum Patrem Jordanum Cardinalem Descriptio Cicladum aliarumq[ue] insularum Foeliciter incipit*, that is, although the maps and handwriting are clearly by Martellus, he does not claim the book as his own creation, instead ascribing it to Buondelmonti—and the text is Buondelmonti's.[31] The manuscript contains many blank folios. The

[27] There is one manuscript of Buondelmonti's *Liber* that contains a world map, namely, Berlin, Staatsbibliothek, Preussischer Kulturbesitz, MS Hamilton 108, f. 81r, probably made about 1470; for discussion see Anna-Dorothee von den Brincken, *Fines terrae: die Enden der Erde und der vierte Kontinent auf mittelalterlichen Weltkarten* (Hannover: Hahnsche Buchhandlung, 1992), pp. 143–145 with plate 47; and my article "Benedetto Cotrugli's Lost *Mappamundi* Found—Three Times," *Imago Mundi* 65.1 (2013), pp. 1–14.

[28] For discussion of *isolarii* as part of the development of the atlas, but without specific reference to Martellus's works, see James Richard Akerman, "On the Shoulders of a Titan: Viewing the World of the Past in Atlas Structure," Ph.D. Dissertation, Pennsylvania State University, 1991, pp. 173–181; also see the very brief reference in Akerman's "From Books with Maps to Books as Maps: The Editor in the Creation of the Atlas Idea," in Joan Winearls, ed., *Editing Early and Historical Atlases* (Toronto: University of Toronto Press, 1995), pp. 3–48, esp. 8. For general discussion of Martellus's *isolario* focusing on the manuscript in the James Ford Bell Library, see Rushika February Hage, "The Island Book of Henricus Martellus," *The Portolan* 56 (2003), pp. 7–23; Nathalie Bouloux, "L'*Insularium Illustratum* d'Henricus Martellus," *The Historical Review—La revue historique* 9 (2012), pp. 77–94, focuses on the manuscript in the Musée Condé in Chantilly.

[29] In the opening of the Minneapolis manuscript, James Ford Bell Library, MS B 1475 fMA, f. [1r], Martellus seems to say that he worked for six years, seeing and touching things himself, in order make sure his book was accurate: see Evelyn Edson, *Cristoforo Buondelmonti: Description of the Aegean and Other Islands, Copied, with Supplementary Material, by Henricus Martellus Germanus* ; (New York: Italica Press, 2018), pp. 19 and 95. But in fact this statement is paraphrased from Christoforo Buondelmonti's isolario, and thus does not reflect Martellus's own efforts. See Christofo Buondelmonti, *Description des îles de l'Archipel grec: version grecque du 'Liber insularum Archipelagi' c. 1420* (Paris: Leroux, 1897), pp. 1 and 157.

[30] The map of Britain is on f. [43v] of the Minneapolis manuscript of Martellus's *Insularium*, and the text is transcribed, not quite correctly, by Hage, "The Island Book of Henricus Martellus" (see note 28 in Chap. 1), p. 15.

[31] For descriptions of the Florence manuscript, see Almagià, "I mappamondi di Enrico Martello" (see note 4 in Front Matter), pp. 295–298, with a comparative chart of some of the maps in the manuscript and other manuscripts of the work on p. 299; and Gentile, *Firenze e la scoperta dell'America* (see note 8 in Chap. 1), pp. 237–240.

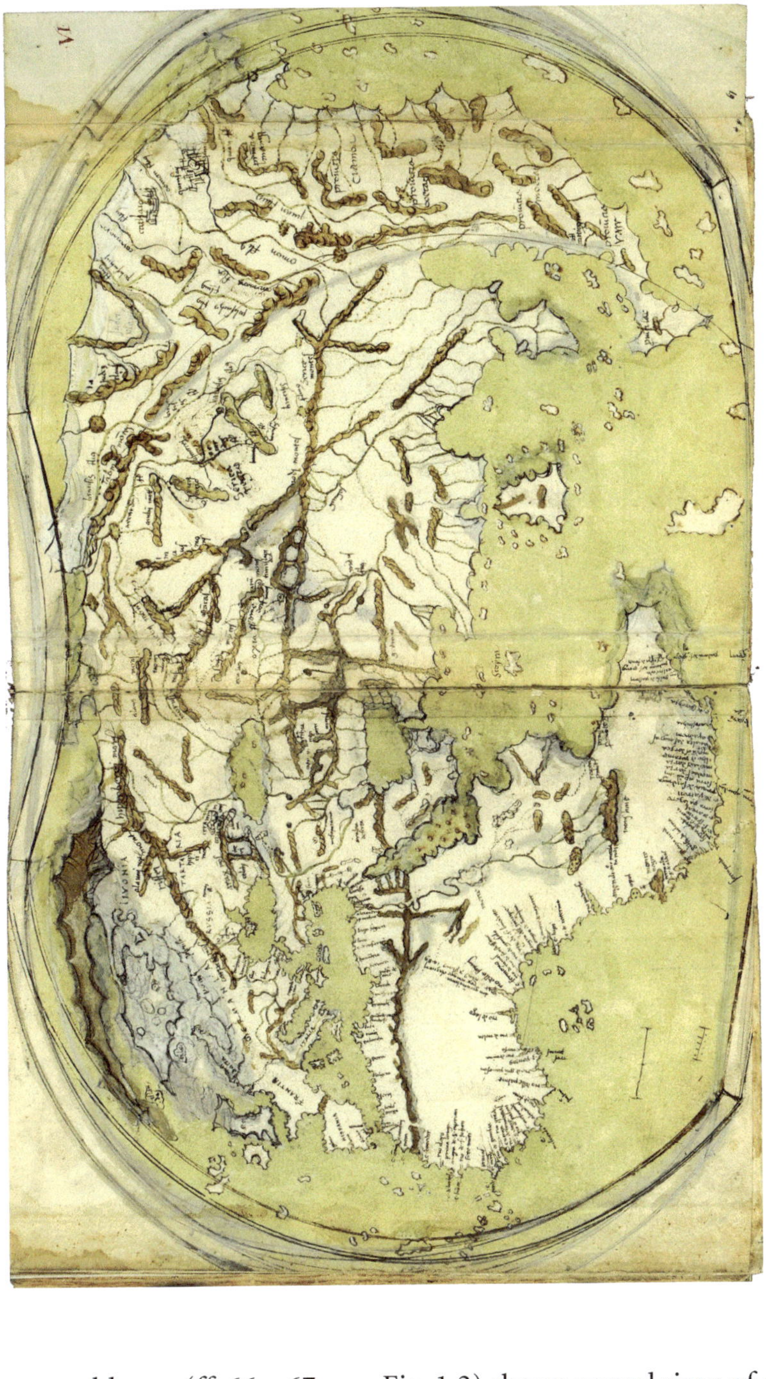

Fig. 1.2 The world map in the Florence manuscript of Martellus's island book (Biblioteca Medicea Laurenziana, Pluteo 29.25, ff. 66v–67r). By concession of the Ministero dei Beni e delle Attività Culturali e del Turismo. (Any further reproduction by any means is prohibited)

world map (ff. 66v–67r, see Fig. 1.2) shows several signs of revision, and on this basis Almagià argued—correctly I believe—that this was Martellus's working manuscript in which he first collected the material that would form his *Insularium*.[32] On the world map, the shape of southern Africa, with its distinctive eastward-jutting peninsula, is

[32] Almagià, "I mappamondi," p. 298. Images of all of the folios of the manuscript are available via http://teca.bmlonline.it/TecaRicerca/index.jsp by searching for the signature "Plut.29.25."

noteworthy. It has not been remarked previously that the presence of the toponym *ilha de fonti* (in other sources *Penedo das Fontes*)[33] in southern Africa entails that the manuscript was made in 1488 or later, for it was that year that Bartolomeu Dias returned from the voyage of discovery in which he named that feature.[34]

The world maps in Martellus's island books include substantially more of the earth's surface than Ptolemaic world maps. They extend further to the south and show that Africa is circumnavigable and also further to the east. Ptolemy's world map covers 180° of longitude and does not include the eastern shore of Asia (see Fig. 1.1), but the world maps in Martellus's island books add information from Marco Polo and reach eastward to include the eastern coast of Asia. These maps do not include a scale of longitude but extend to about 240° east, that is, about 60° further than Ptolemy (see below for this calculation).

2. Minneapolis, James Ford Bell Library, MS B 1475 fMA. The manuscript consists of 43 folios with 85 maps.[35] The text begins *Cristofori Ensenii Florentini sacerdotis dignissimi Descriptio arcipelagi et Cicladum alia[rumque] insula[rum] foeliciter incipit*, that is, the incipit is very similar to that in the Florence manuscript, and again the text is Buondelmonti's—and thus very similar to that in the Florence manuscript. The Florence manuscript, however, contains a much more elaborate cartographic program, particularly relating to areas outside the Mediterranean, and includes a world map, while the Minneapolis manuscript does not have a world map. Given the signs that the Florence manuscript was Martellus's working copy, it seems likely that the Minneapolis manuscript is the later of the two but was a less elaborate commission. The date of c. 1475 that is assigned to the Minneapolis

[33] For discussion of the identification of this toponym, see E. H. L. Schwarz, "Bartholomieu Dias's Furthest East," *South African Journal of Science* 9 (1912), pp. 103–107.

[34] On Bartolomeu Dias and the *ilha de fonti,* see Eric Axelson, "The Dias Voyage, 1487–1488: Toponymy and Padrões," *Revista da Universidade de Coimbra* 34 (1988), pp. 29–55; also published as an offprint by the Centro de Estudos de História e Cartografia Antiga, Serie Separatas, no. 189 (1988), esp. pp. 43–44; and W. G. L. Randles, *Bartolomeu Dias and the Discovery of the South-East Passage Linking the Atlantic to the Indian Ocean* (1488), *Revista da Universidade de Coimbra* 34 (1988), pp. 19–28, also published as an offprint by the Centro de Estudos de História e Cartografia Antiga, Série separatas 188 (1988), esp. pp. 25–26; and reprinted in W. G. L. Randles, *Geography, Cartography and Nautical Science in the Renaissance: The Impact of the Great Discoveries* (Aldershot, Hampshire, Great Britain; and Burlington, VT: Ashgate/Variorum, 2000).

[35] For descriptions of the Minneapolis manuscript, see Christopher U. Faye and William H. Bond, *Supplement to the Census of Medieval and Renaissance Manuscripts in the United States and Canada* (New York: Bibliographical Society of America, 1962), p. 299, no. 4; and Paul Oskar Kristeller, *Iter Italicum: A Finding List of Uncatalogued or Incompletely Catalogued Humanistic Manuscripts of the Renaissance in Italian and Other Libraries* (London: Warburg Institute, 1963–1997), vol. 5, p. 274b. This manuscript was not known to Almagià and thus is not included in his survey of Martellus's works cited above; the fullest but somewhat general account of the manuscript is Rushika February Hage, "The Island Book of Henricus Martellus," *The Portolan* 56 (2003), pp. 7–23, but unfortunately her transcriptions and translations of selections of the Latin text are not all accurate. Evelyn Edson has now published an edition and translation of the text of the Minneapolis manuscript, titled *Cristoforo Buondelmonti: Description of the Aegean and Other Islands, Copied, with Supplementary Material, by Henricus Martellus Germanus* (New York: Italica Press, 2018).

manuscript in the brief catalog descriptions seems to be without any firm foundation, and if the Minneapolis manuscript was in fact made after the Florence one, it would have to be dated 1488 or later.

3. Leiden, Leiden University Library, MS VLF 23. This would seem to be the earliest surviving manuscript of Martellus's work in which he asserts his authorship and gives it the title it bears in later manuscripts, *Insularium*. The manuscript consists of 80 folios, with the world map on ff. 65v–66r (Fig. 1.3).[36] The manuscript has suffered from water damage, and the world map is in a poor state of preservation: the blue paint for the oceans has flaked off in several places, and much of the paint originally applied to the mountains now adheres to the opposite folio.[37] The depiction of southern Africa is different in the Leiden world map than it is in the Florence one: on the Florence map, the continent ends in the south in an eastward-jutting peninsula, while on the Leiden map, the coastline is not indicated between *ilha de fonte* on the southern coast and *prassum promontorium* on the southeastern coast.[38] Again, the presence of the toponym *ilha de fonti* on the Leiden map indicates that the manuscript was made in 1488 or later. Preceding and following the world map, that is to say on ff. 64v and 66r, there are much the same excerpts from Chapters 1 and 2 of Enea Silvio Piccolomini's *Historia rerum ubique gestarum* that appear with the world map in Martellus's manuscript of Ptolemy's *Geography* which is in Florence, Biblioteca Nazionale Centrale, Magliabechiano XIII 16 (discussed above).
4. Chantilly, Bibliothèque du Musée Condé, MS 698 (483). This is a luxury manuscript, elaborately decorated, of 78 folios, which bears the title *Insularium illustratum Henrici Martelli Germani*.[39] Meurgey de Tupigny

[36] For descriptions of the Leiden manuscript, see Almagià, "I mappamondi di Enrico Martello" (see note 4 in Front Matter), pp. 291–292, with a comparative chart of some of the maps in the manuscript on p. 299; K. A. de Meyïer and P. F. J. Obbema, *Codices Vossiani Latini* (Leiden: Universitaire pers. Leiden, 1973–1984) (= Bibliotheca Universitatis Leidensis. Codices manuscripti. XIII–XVI), vol. 1, pp. 47–49.

[37] The world map in the Leiden manuscript is discussed by K. A. Kalkwiek, "Three MappaeMundi from the University Library in Leyden," *Janus* 62.1–3, (1975), pp. 17–41, esp. 34–39.

[38] Although in his 1940 article "I mappamondi di Enrico Martello" (see note 4 in Front Matter), Almagià argued that the Florence manuscript was that in which Martellus first put together his *Insularium*, and the world map in that manuscript was clearly revised by Martellus, in a later article he asserts, without offering any supporting argument, that the map in the Leiden manuscript is the oldest: see Roberto Almagià, "L'evoluzione delle conoscenze sulla figura dell'Africa e sull'Oceano Indiano secondo alcune carte italiane dei secoli XV e XVI," in *Comptes Rendus du Congrès International de Géographie, Lisbonne 1949* (Lisbon: Union Géographique International, 1950–1952), vol. 4, pp. 219–224, at 221. Perhaps this suggestion was based on the incomplete coastline of southern Africa in the Leiden world map; in any case, I do not see any strong reason to think that the Leiden world map is earlier than the Florence one.

[39] Descriptions of the manuscript include Almagià, "I mappamondi di Enrico Martello" (see note 4 in Front Matter), pp. 294–295, with a comparative chart of some of the maps in the manuscript on p. 299; Jacques Meurgey de Tupigny, *Les principaux manuscrits à peintures du Musée Condé à Chantilly* (Paris: Pour les membres de la Société Française de Reproductions de Manuscrits à Peintures, 1930), pp. 192–194 and plate 129; and Albinia de la Mare, "The Florentine Scribes of Cardinal Giovanni of Aragon," in Cesare Questa and Renato Raffaelli, eds., *Il Libro e il testo: atti del convegno internazionale, Urbino, 20–23 settembre 1982* (Urbino: Università degli studi di Urbino,

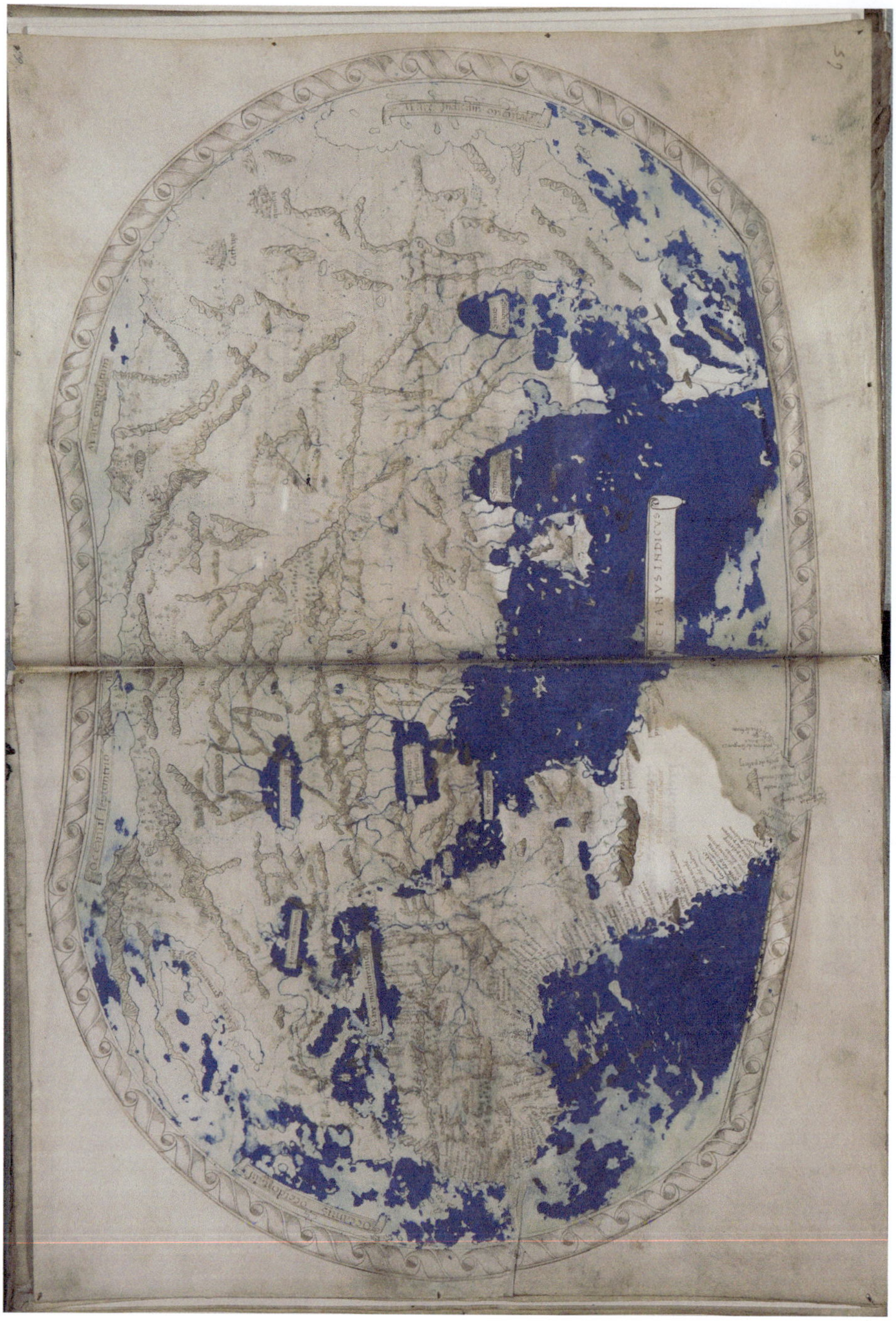

Fig. 1.3 The world map in the Leiden manuscript of Martellus's Insularium *(Leiden University Library, MS VLF 23, ff. 65v–66r). (Courtesy of Leiden University Library)*

suggests that the client who commissioned the manuscript was a prince of England on the basis of a phrase in the text describing Ireland, and the author does address the reader in the vocative and refer to the reader's grandfather King Henry,[40] but this phrasing does not indicate the

1984), pp. 243–293, esp. 278, no. 37. For a more detailed account, see Bouloux, "L'*Insularium Illustratum* d'Henricus Martellus" (see note 28 in Chap. 1).

[40]Meurgey de Tupigny, *Les principaux manuscrits* (see note 39 in Chap. 1), pp. 193–194, transcribes the phrase: *...usque ad tempora illustrissimi regis Anglorum Henrici avi tui, princeps sacratissime, qui primis, expulsis obsenis Hiberniensibus gentibus, terram Anglis possidendam [f]oedis militaribus distinxit....*

manuscript's patron. Rather, Martellus has borrowed his description of Ireland from Gervase of Tilbury's *Otia imperialia*, where this same phrasing occurs[41]: Gervase dedicated his book to the Holy Roman Emperor Otto IV, who was the grandson of King Henry II of England. The two folios in the Chantilly manuscript which had been ff. 75–76 are missing, and these contained the world map: on f. 74v there is the beginning of a text titled "Mundi forma" that introduced the map, and this same text appears with the world map in the London manuscript of the *Insularium* and also in the Florence manuscript of Ptolemy's *Geography* which was made by Martellus (Biblioteca Nazionale Centrale, Magliabechiano XIII 16). The text "Mundi forma" is copied almost verbatim from Chapters 1 and 2 of the *Historia rerum ubique gestarum* of Enea Silvio Piccolomini (later Pope Pius II), which was first printed in 1477; this text addresses the question of the shape of the world and also whether the *orbis terrarum* (the inhabited part of the world) is circumnavigable. The latter part of the manuscript contains various regional maps and maps of islands outside of the Mediterranean: on f. 52, of Britain and Ireland; on f. 55, of the Holy Land; on f. 57, of Italy; on f. 59, of Spain; on f. 61, of France; on f. 63, of Germany; on f. 66, of northwestern Europe including Iceland and Greenland; on f. 67, Greece; on f. 69, Asia Minor; on f. 72, Taprobana; and then, following the missing folios with the world map, a nautical chart of the western coasts of Europe and North Africa, a chart of the Mediterranean, and a chart of the Black Sea. Again in this case, we see Martellus giving the *isolario* some of the character of an atlas. The loss of the world map is unfortunate, not least because it deprives us of what might have been a valuable tool for dating the manuscript in relation to Martellus's others. But the title alone tells us that this was not one of the earliest manuscripts of the work.

5. London, British Library, Add. MS 15760. This is the best known manuscript of Martellus's *Insularium*, whose world map (ff. 68v–69r) is the most elaborate, important, and widely reproduced (Fig. 1.4).[42] As in the Chantilly manuscript, the world map is introduced by text (ff. 65r–68r) from Chapters 1 and 2 of *Historia rerum ubique gestarum* of Enea Silvio

[41] See Gervase of Tilbury, *Otia imperialia: Recreation for an Emperor*, ed. and trans. S. E. Banks and J. W. Binns (Oxford: Clarendon Press, 2002), pp. 308 (Latin) and 309 (English).

[42] There is a very brief description of BL Add. MS 15760 in *Catalogue of Additions to the Manuscripts in the British Museum in the Years MDCCCXLVI-MDCCCXLVII* (London: Printed by Order of the Trustees, 1864), p. 26; for more detailed descriptions, see John Holmes and Frederic Madden, *Catalogue of the Manuscript Maps, Charts, and Plans, and of the Topographical Drawings in the British Museum* (London: Printed by Order of the Trustees, 1844), vol. 3, pp. 32–34, which includes lists of the maps in the manuscript (this same information may be found in the BL's online catalogue); and Almagià, "I mappamondi di Enrico Martello" (see note 4 in Front Matter), p. 292, with a comparative chart of some of the maps in the manuscript on p. 299, and discussion of the world map on pp. 304–306. The best reproductions of the world map are those in Kenneth Nebenzahl, *Atlas of Columbus and the Great Discoveries* (Chicago: Rand McNally, 1990), pp. 15–17; and Guglielmo Cavallo, ed., *Cristoforo Colombo e l'apertura degli spazi: Mostra storico-cartografica* (Rome: Istituto Poligrafico e Zecca dello Stato, 1992), vol. 1, pp. 420–421.

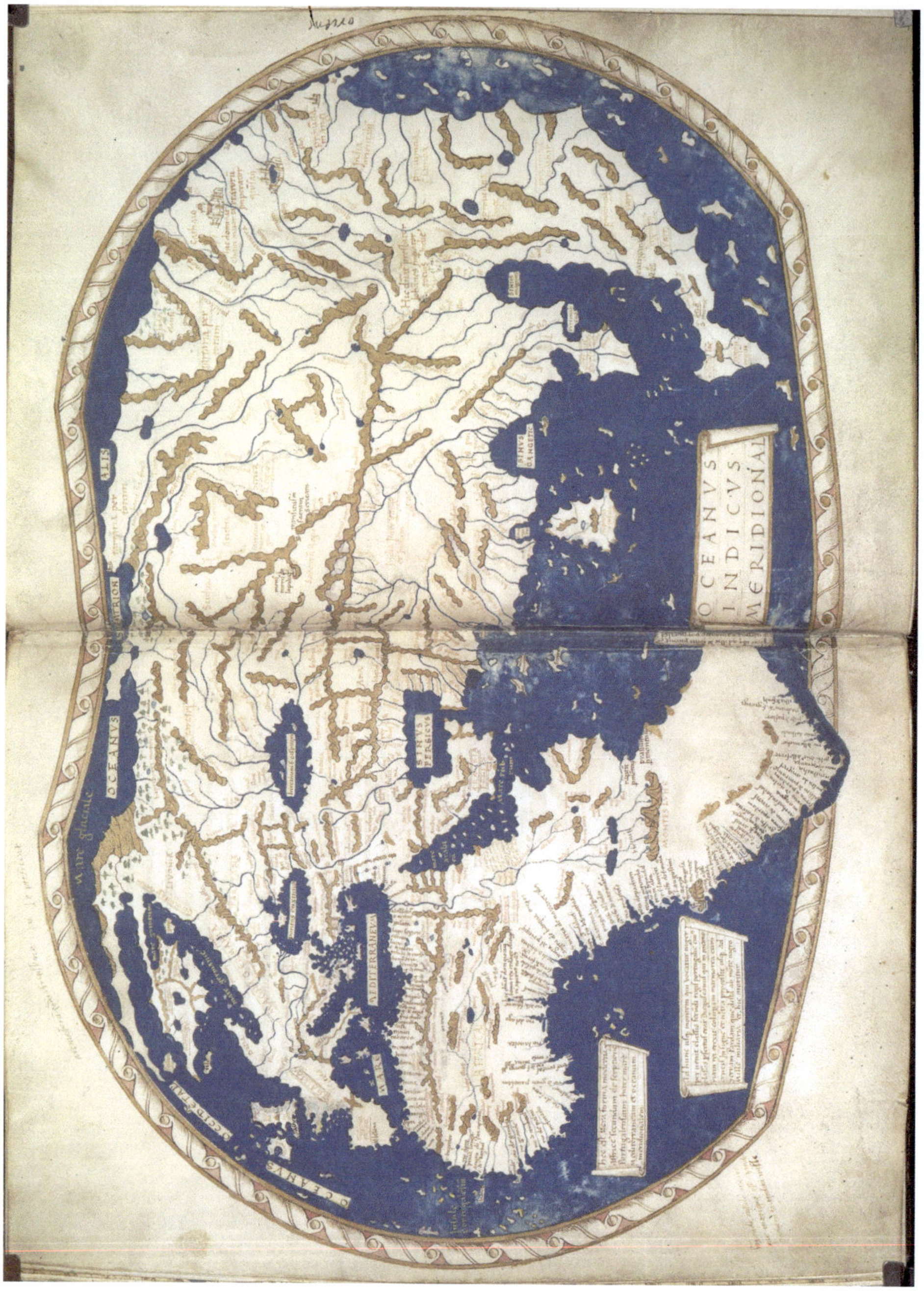

Fig. 1.4 The world map in the London manuscript of Martellus's Insularium. *(British Library, Add. MS 15760, ff. 68v–69r, © The British Library Board)*

Piccolomini. The world map is of particular interest partly because it (like the rest of the manuscript) is in very good condition, and it contains three legends relating to Portuguese exploration of Africa that demonstrate Martellus's interest in using the most recent and accurate sources available to him. The first reads:

hec est uera forma moderna affrice secundum descr[i]ptionem Portugalensium Inter mare Mediterraneaum et oceanum meridionalem.

This is the modern true shape of Africa according to the map of the Portuguese, from the Mediterranean Sea to the southern ocean.

In proclaiming the superiority of recent Portuguese maps, Martellus is implicitly criticizing the Ptolemaic image of the continent.[43] The second legend reads:

> *Ad hunc usque montem qui vocatur niger peruenit classis secundi regis portugalie cuius classis prefectus erat Diegus canus qui in memoriam rei erexit columnam marmoream cum crucis Insigne' et ultra processit usque ad Serram Pardam que distat ab monte nigro Mille miliaria et hic moritur.*
>
> *This mountain, which is called Monte Negro, was reached by the fleet of [John] the second King of Portugal; the fleet was commanded by Diogo Cão, who to commemorate the fact, set up a prominent marble column with a cross on it, and proceeded to the Serra Parda [Ponta dos Farilhões, at 22°10'S] which is 1000 miles from Monte Negro, and here he died.*[44]

This legend refers to Diogo Cão's second voyage of 1484–1486[45]; the column or *padrão* that Cão erected on Monte Negro is prominently illustrated on all three of Martellus's world maps in manuscripts of his *Insularium* (Florence, Leiden, and London) and also on Martin Waldseemüller's 1507 map (at *Caput nigrum*); and the *padrão* itself was discovered in the late nineteenth century.[46] The third legend, which is at the southeastern tip of Africa, reads:

---

[43] For discussion of Martellus's revision of the shape of Africa, see Francesc Relaño, *The Shaping of Africa: Cosmographic Discourse and Cartographic Science in Late Medieval and Early Modern Europe* (Burlington, VT: Ashgate, 2002), pp. 165–168; Armando Cortesão, *History of Portuguese Cartography* (Lisbon: Junta de Investigações do Ultramar, 1969–1971), vol. 2, p. 204; and Randles, "Bartolomeu Dias and the Discovery of the South-East Passage" (see note 34 in Chap. 1), p. 24. Incidentally Martellus's criticism of Ptolemy, though implicit, is still substantially more aggressive than the very deferential attitude of Donnus Nicolaus Germanus, with whom Martellus had worked, toward revising Ptolemy, as expressed in the dedication that appears in some manuscripts of Ptolemy that Nicolaus made. This dedication is edited in Joseph Fischer, *The Discoveries of the Norsemen in America: With Special Relation to their Early Cartographical Representation* (London: Henry Stevens, Son & Stiles; St. Louis: B. Herder, 1903), pp. 112–118, esp. 112–113, and is translated into English in *Geography of Claudius Ptolemy*, trans. Edward Luther Stevenson (New York: New York Public Library, 1932), pp. 19–24, esp. 19–20.

[44] There has been considerable debate about whether the phrase *et hic moritur* refers to the death of Diogo Cão; this is certainly the literal meaning of the Latin. For discussion see Augusto César da Silva Castro Júnior, "Diogo Cão e a legenda de Henrique Martelo," *Actas do Congresso Internacional de História dos Descobrimentos* (Lisbon: Comissão Executiva das Comemorações do V Centenário da Morte do Infante D. Henrique, 1961), vol. 2, pp. 85–109; and Américo da Costa Ramalho, "Sobre a data da morte de Diogo Cão," in his *Estudos sobre a época do Renascimento* (Coimbra: Instituto de Alta Cultura, 1969; Lisbon: Fundação Calouste Gulbenkian and Junta Nacional de Investigação Científica e Technológica, 1997), pp. 2–8. One possibility would be to emend "moritur" to "moratur," indicating that the fleet stayed at the location of the column for a time. W. G. L. Randles in "The Atlantic in European Cartography and Culture from the Middle Ages to the Renaissance," in *Geography, Cartography and Nautical Science in the Renaissance: The Impact of the Great Discoveries* (Aldershot, Hampshire, Great Britain, and Burlington, VT: Ashgate/Variorum, 2000), p. 18, takes the fleet as the subject of "moritur" and translates the verb as "went no further," but this is to stretch the meaning of the Latin too far.

[45] For discussion of Cão's second voyage, see E. G. Ravenstein, "The Voyages of Diogo Cão and Bartholomeu Dias, 1482–1488," *Geographical Journal* 16.6 (1900), pp. 625–655, esp. 633–638.

[46] On the discovery of the *padrão*, see Luciano Cordeiro, "O ultimo padrão de Diogo Cão," *Boletim da Sociedade de Geografia de Lisboa* 14.11 (1895), pp. 885–894. For other discussions of this legend on the world map in the London manuscript of Martellus's *Insularium*, see Cortesão, *History of Portuguese Cartography* (see note 43 in Chap. 1), vol. 2, p. 204; Axelson, "The Dias Voyage" (see note 34 in Chap. 1), p. 32; and Randles, "Bartolomeu Dias and the Discovery of the South-East Passage" (see note 34 in Chap. 1), p. 22.

huc usque ad ilhe de fonti peruenit ultima navigatio portugalensium anno domini 1489.
To this place, the Island of the Fountain, the latest voyage of the Portuguese reached in the year of the Lord 1489.

This legend refers to the voyage of Bartolomeu Dias in 1487–1488, who reached the *ilha de fonti*. It has been suggested that the reference to 1489 indicates the date when the map was made rather than being a mistake for the year of Dias's voyage; however, to me, it seems more likely that 1489 is a mistaken indication of the year of Dias's voyage.

In any case, these legends on the world map in the London manuscript of Martellus's *Insularium* clearly attest to his interest in accurate cartographic data and the latest discoveries. Given that the Florence, Leiden, and London manuscripts were all certainly made between 1488 (the date of the discovery of the *ilha de fonti*) and the 1490s (when knowledge of Columbus's first voyage began to be diffused, of which the manuscripts betray no knowledge), it seems that this was a period of intense cartographic activity for Martellus.

6. Bern, Burgerbibliothek, MS 144/2, i.e., the second work in MS 144. There is no published catalog description of this manuscript. It contains on ff. 198r–198v the introduction and list of the islands discussed in Martellus's *Insularium* but none of the text proper and none of the maps.[47] Paleographic analysis indicates that the folio is from the second half of the sixteenth century.[48]

SEPARATE WORLD MAPS

1. A world map by Martellus, now lost, depicted in a miniature in vol. 7 of the Bíblia dos Jerónimos, one of the finest Bibles of the Italian Renaissance.[49] The Bible was made for the Duke of Beja, who during the 3-year process of making of the manuscript became King Manuel of

[47] I thank Renate Burri for her help with this manuscript.

[48] I thank Ilya Dines and Maddalena Signorini for their help with the paleographic analysis.

[49] The shelf mark of this volume of the Bible is Lisbon, Arquivos Nacionais da Torre do Tombo, MS 161/7, the final "7" indicating volume 7. For discussion of the Bíblia dos Jerónimos see Annarosa Garzelli, ed., *Miniatura fiorentina del Rinascimento 1440–1525: un primo censimento* (Florence: Giunta regionale toscana and La Nuova Italia, 1985), vol. 1, pp. 233–235, and vol. 2, Figs. 821–834 and 917–919; Jonathan J. G. Alexander, ed., *The Painted Page: Italian Renaissance Book Illumination, 1450–1550* (Munich and New York: Prestel, 1994), pp. 49–51; Albinia C. de la Mare, "Notes on Portuguese Patrons of the Florentine Book Trade in the Fifteenth Century," in Katherine J. P. Lowe, ed., *Cultural Links between Portugal and Italy in the Renaissance* (Oxford: Oxford University Press, 2000), pp. 167–181; Martim de Albuquerque and Arnaldo Pinto Cardoso, *A Bíblia dos Jerónimos* (Lisbon: Bertrand Editora, 2004); and Arnaldo Pinto Cardoso, "The Bíblia dos Jerónimos: From Florence to Lisbon," photography by Massimo Listri, trans. Judith Landry, *FMR* 10 (Dec. 2005/Jan. 2006), pp. 43–72.

Fig. 1.5 A miniature in the Bíblia dos Jerónimos by Attavante degli Attavanti showing a world map by Martellus in the background (Lisbon, Arquivos Nacionais da Torre do Tombo, MS 161/7, f. 2r). (Courtesy of the Torre do Tombo)

Portugal, and was painted by Attavante degli Attavanti.[50] Volume 7 of the Bible is dated July, 1497, and the miniature in question is on f. 2r. It shows St. Jerome writing at a lectern, and on the wall in the background, there is a shelf of books, a clock, and a wall map. There can be little doubt that the map depicted (Fig. 1.5) is the work of Martellus: the similarity with the world maps in manuscripts of his *Insularium illustratum* is overwhelming.[51] The projection is the same, the area of the world depicted—from the eastern Atlantic to the eastern coast of Asia—is the same, and the Asian peninsula jutting southwest into the Indian Ocean is the same. And yet the artist has made a significant adjustment so that the viewers of the miniature will be able to easily recognize it as a map: he has made Europe and the Mediterranean, the geographic features most readily recognizable for European viewers, much larger than they are on Martellus's maps. On the world map in the British Library man-

[50]The contract for the painting of the manuscript, which is dated April 23, 1494, still survives. The text of the contract is supplied by Gaetano Milanesi, *Nuovi documenti per la storia dell'arte Toscana dal XII al XV secolo* (Florence: G. Dotti, 1901; Soest: Davaco Publishers, 1973), pp. 164–166; reprinted in Jonathan J. G. Alexander, *Medieval Illuminators and their Methods of Work* (New Haven and London: Yale University Press, 1992), pp. 181–182; and there is an English translation by Dario Tessicini in Carol M. Richardson, "The Contract Between Attavante and a Florentine Merchant for an Illuminated Manuscript," in Carol M. Richardson, Kim W. Woods and Michael W. Franklin, eds., *Renaissance Art Reconsidered: An Anthology of Primary Sources* (Oxford: Blackwell Publishing, 2007), pp. 329–332.

[51]For a more detailed discussion of this map, see Chet Van Duzer, "Graphic Record of a Lost Wall Map of the World (c. 1490) by Henricus Martellus," *Peregrinations: Journal of Medieval Art & Architecture* 5.2 (2015), pp. 48–64.

uscript of Martellus's *Insularium illustratum* (see Fig. 1.4), the Mediterranean extends about halfway from the western edge of Europe toward the centerline of the map, while in Attavante's image, it extends all the way to the map's centerline and thus is fully twice as large as it should be.

It is possible to situate the map within the chronological sequence of Martellus's cartographic works with some confidence. On most of Martellus's world maps, the southern tip of Africa projects beyond the southern border of the map (see Fig. 1.4), but it does not do so on the map depicted in the Bíblia dos Jerónimos, and it does not in one of Martellus's earliest world maps, that in the Florence manuscript of his *Insularium* (see Fig. 1.2). In this manuscript, as indicated above, the world map shows several signs of revision, suggesting that Martellus was still working out how he wanted to depict the world. This feature of the map in the Bíblia dos Jerónimos thus associates it with Martellus's early cartographic work. Another feature of the map that associates it with Martellus's early work is its lack of the additional geographical data in the east—the expanse of the proto-Pacific and Japan—that Martellus includes on his Yale map.

The proportions of the map are different from those of the world maps in manuscripts of Martellus's *Insularium*—specifically, the map is wider—which suggests that it was not a bifolium from the *Insularium* adapted for wall display but rather designed as a wall map from the start. In addition, the map shows an island in the northern ocean that only appears on only one other map by Martellus, namely, his wall map at Yale.[52] The presence of this island tends to confirm that the map depicted by Attavante was designed as a wall map. Thus Martellus was evidently making wall maps from an early point in his career as a cartographer.

2. Florence, Biblioteca Nazionale Centrale, Landau Finaly, Carte Rosselli, planisfero. This undated world map was printed by Francescso Rosselli, a very active engraver, manuscript painter,[53] and cartographer who was also one of the first map sellers.[54] The map measures 46 × 28.5 cm

[52] See the discussion of *bannona insula septentrionalis* on the Yale Martellus map below.

[53] On Rosselli's work as an engraver and manuscript painter, see Mirella Levi d'Ancona, "Francesco Rosselli," *Commentari* 16 (1965), pp. 56–76; Annarosa Garzelli, "Il contributo di Francesco Rosselli al codice miniato," in *Miniatura fiorentina del rinascimento, 1440–1525: un primo censimento* (Florence: Giunta Regionale Toscana e La Nuova Italia, 1985), vol. 1, pp. 173–188; Suzanne Boorsch, "Francesco Rosselli," in Arthur R. Blumenthal, ed., *Cosimo Rosselli, Painter of the Sistine Chapel* (Winter Park, Florida: Cornell Fine Arts Museum, Rollins College, 2001), pp. 208–244; and Suzanne Boorsch, "The Case for Francesco Rosselli as the Engraver of Berlinghieri's *Geographia*," *Imago Mundi* 56.2 (2004), pp. 152–169.

[54] On Rosselli's work as a cartographer, see Sebastiano Crinò, "I planisferi di Francesco Rosselli dell'epoca delle grandi scoperte geografiche," *La Bibliofilía* 41 (1939), pp. 381–405, esp. 393–401 on the map under discussion here; Roberto Almagià, "On the Cartographic Work of Francesco Rosselli," *Imago Mundi* 8 (1951), p. 27–34, esp. 31–32; and Tony Campbell, *The Earliest Printed Maps, 1472–1500* (London: British Library, 1987), pp. 70–78, esp. 72–74. The Martellus-Rosselli planisphere is also discussed and illustrated in Cavallo, *Cristoforo Colombo e l'apertura degli spazi* (see note 42 in Chap. 1), vol. 1, pp. 521–524; and in Rodney W. Shirley, *The Mapping of the World: Early Printed World Maps, 1472–1700* (London: Holland Press, 1983), #18, pp. 16–17.

Fig. 1.6 A world map evidently by Martellus printed by Francesco Rosselli (Florence, Biblioteca Nazionale Centrale, Landau Finaly, Carte Rosselli, planisfero). By concession of the Ministero dei Beni e delle Attività Culturali e del Turismo/Biblioteca Nazionale Centrale, Firenze. (Further reproduction in any medium is forbidden)

(Fig. 1.6). Although it was printed by Rosselli and there is no explicit connection between Martellus and the map, its projection, layout, visual style, and geography, particularly the shape of Africa and of the "Fourth Peninsula" in southeastern Asia, make the map's dependence on Martellus unmistakable—and given the strong geographical differences from Rosselli's other maps, we may be certain that the design of the map came from Martellus. The map has more toponyms and legends than those in the manuscripts of his *Insularium* (several of which legends will be examined below), and also includes a line indicating the equator, which most of Martellus's manuscript maps, except for that in the Florence manuscript of his island book, and his Yale map lack. The Martellus-Rosselli map also has wind-heads decorating the margins of the map, no doubt inspired by those in a world map in a manuscript of Ptolemy's *Geography*.[55]

In the South Atlantic off the west coast of Africa there is a large banner—in the same position as one of the banners on Martellus's world map in the London manuscript of his *Insularium*—with a legend that may be considered the map's title:

> *forma uniuersalis totius quod oceano mari ambitur cum superioris Indie portione post ptholomej tempus inuenta cumque ea parte Aphrice quam temporibus nostris lusitanj naute per lustrarunt ita se habet*
>
> *The complete form of the whole [world] which is surrounded by the ocean sea, with the part of upper India that was discovered after the time of Ptolemy, and the part of Africa which Portuguese sailors explored in our times.*[56]

Again Martellus shows his pride in his knowledge of recent geographical discoveries. The indication here that the world is surrounded by the ocean recalls Martellus's quotation from Chapter 2 of Enea Silvio Piccolomini's *Historia rerum ubique gestarum* in his *Insularium illustratum*: evidently the question of the circumnavigability of the *orbis terrarum* was of considerable interest to Martellus. There is another banner with a legend near the southeasternmost part of Africa, which addresses the Portuguese voyages of exploration and runs thus:

> *Huc usque ad ultimam columnam pervenerunt nautae lusitani. 1498.*
>
> *To this point, to the last column, came the Portuguese sailors. 1498.*

This map, like the world maps in the Florence and Leiden manuscripts of Martellus's *Insularium*, includes an image of this *padrão*, the furthest set up by Bartolomeu Dias, which he erected at Kwaaihoek on March 12,

[55] There are some brief remarks about the wind heads on the Martellus-Rosselli map in Chet Van Duzer, "A Newly Discovered Fourth Exemplar of Francesco Rosselli's Oval Planisphere of c.1508," *Imago Mundi* 60.2 (2008), pp. 195–201, esp. 198–199.

[56] The English translation is based on that by Edson, *The World Map, 1300–1492* (see note 7 in Chap. 1), p. 220.

1488[57]; the London map does not have an image of it, probably for lack of space west of the *ilha de fontj*, but does have the toponym (*padram de s. giorgi*). The date in the legend just cited is certainly an error for 1489, the date that Martellus gave on his world map in the London manuscript of his *Insularium*.[58] With regard to the date of the map, certainly it is between 1489 (when some report of Bartolomeu Dias's voyage might have been diffused) and the mid-1490s (when knowledge of Columbus's first voyage began to be well diffused). So we can dismiss Crinò's suggestion that the map should be dated between 1498 and 1500.[59] Tony Campbell has suggested a date of 1492,[60] but given the apparent lack of data that would permit us to assign it to a specific year, perhaps it is more important to try to establish its place in the sequence of other maps by Martellus.

One way we can try to establish this place is to consider Martellus's handling of the southern tip of Africa in relation to the map's border. On the world map in the Florence manuscript of Martellus's island book, which is probably the earliest, Africa does not reach the lower border of the map (Fig. 1.2); in the Leiden manuscript, Africa goes beyond the border, and it seems that the border was partly erased by the cartographer, and then the southern tip of Africa drawn over it (Fig. 1.3). Both the shape of the southern tip of Africa and its handling by Martellus in the Martellus-Rosselli map is extremely similar: it is drawn over the border and extends beyond it (Fig. 1.6). In the London manuscript, the border is "broken" to make room for southern Africa, and not only the southern tip of Africa but also the surrounding water extends beyond the border (Fig. 1.4). As we will see below, Martellus's handling of southern Africa in his map at Yale, which seems to have been made after the maps in the surviving manuscripts of his *Insularium*, is very similar to what we see in the London map. Thus the Martellus-Rosselli map is closest in design (with respect to southern Africa) and also probably in date to the map in the Leiden manuscript of the *Insularium*, and so it is probably later than the Florence map and earlier than the London map.

The Martellus-Rosselli has a number of descriptive legends, and some of these read like abbreviated versions of those on the Yale Martellus map.

[57]On this *padrão* see Eric Axelson, "Discovery of the Farthest Pillar Erected by Bartholomew Dias," *South African Journal of Science* 35 (1938), pp. 417–429; and Axelson, "The Dias Voyage" (see note 34 in Chap. 1), pp. 47–48.

[58]So Relaño, *The Shaping of Africa* (see note 43 in Chap. 1), p. 174. Campbell, *The Earliest Printed Maps* (see note 54 in Chap. 1), p. 73, suggests that the map's 1498 is an error for 1488, but see the next note.

[59]Crinò, "I planisferi di Francesco Rosselli" (see note 54 in Chap. 1), p. 394, says that the date 1498 is correct and asserts that the map was made between 1498 and 1500, but we have no other evidence of Martellus's cartographic work that late in the fifteenth century. Almagià, "On the Cartographic Work" (see note 54 in Chap. 1), p. 31, insists that the correct reading of the date is 1488, and that this is the date of the map, rather than the date of Bartolomeo Dias's voyage. But to me the reading of 1498 is quite clear, and 1488 would be impossible as a date for the map, as Dias returned to Lisbon in December of 1488.

[60]Campbell, *The Earliest Printed Maps* (see note 54 in Chap. 1), pp. x, 70, and 74.

For example, north of the Caspian Sea the Martellus-Rosselli map has a legend that reads *Hic reperietur abundantia omnium varorum*, "Here there is found an abundance of all squirrels." The Yale Martellus map has a similar but longer legend in the same location, which reads *hic magna habundantia varorum armelinorum* [ ]*rorum vulpium cebelinorum*, "Here there is a great abundance of squirrels, ermines, [ ], foxes, and sables." Unfortunately there does not seem to be a way to determine which map came first, the Yale Martellus map or the Martellus-Rosselli map.

3. New Haven, Beinecke Rare Book and Manuscript Library, Art Store 1980.157 (on permanent display) (Fig. 1). Here I offer a discussion of the general characteristics of this map, which will be followed below by a transcription, translation, and analysis of many of the legends on the map. The manuscripts of Martellus's *Insularium* testify to his ability to secure commissions from wealthy clients, and his world map at Yale provides further evidence of this ability, for it is indeed a luxury production. Its great size, 122 × 201 cm, or about 4 × 6.5 feet; extensive hand painting, including the writing of many detailed legends; and the use of gold in some lettering, all bespeak an affluent client, but unfortunately the map provides no clue as to that client's identity. The map is signed along its southern border *Opus Henrici Martelli Germani*,[61] "the work of Henricus Martellus Germanus," but does not bear a date.

The monumental size and elaborate hand painting of the Yale Martellus map make it part of the medieval and Renaissance tradition of large maps intended for display by nobles and royalty as symbols of power, prestige, and culture.[62] It is possible that there is also a very specific humanist consideration behind the map's dimensions. We know that Martellus was familiar with the work of the Greek geographer Strabo,[63] as he cites him in the text block in the lower left corner of the map (see below), and Strabo in his *Geography* recommends that if one is going to make a terrestrial

[61] R. A. Skelton in his description of the map in Destombes's *Mappemondes*, p. 229, mistakenly read the signature as *Opus Henricus Martellus Germanus*, and this mistaken reading has been repeated by several authors.

[62] On the display of medieval and Renaissance maps in noble settings, see Mark S. Rosen, "Maps as Decoration in Medieval Europe" and "Maps as Decoration in the Fifteenth and Sixteenth Centuries" in his "The Cosmos in the Palace: The Palazzo Vecchio Guardaroba and the Culture of Cartography in Early Modern Florence, 1563–1589," Ph.D. Dissertation, University of California at Berkeley, 2004, pp. 57–69 and 90–122, respectively. When the Yale Martellus map was made, the market for maps in Italy was about to be immensely broadened, so that smaller maps became common decorative pieces in more humble houses: see Federica Ambrosini, "'Descrittioni del mondo' nelle case venete dei secoli XVI e XVII," *Archivio Veneto*, series 5, vol. 117, no. 152 (1981), pp. 67–79; Genevieve Carlton, "Making an Impression: The Display of Maps in Sixteenth-Century Venetian Homes," *Imago Mundi* 64.1 (2012), pp. 28–40; and Genevieve Carlton, *Worldly Consumers: The Demand for Maps in Renaissance Italy* (Chicago: University of Chicago Press, 2015).

[63] On the Renaissance reception of Strabo, see Milton Vasil Anastos, "Pletho, Strabo, and Columbus," *Annuaire de l'Institut de Philologie ed. d'Histoire Orientales et Slaves* 12 (1952), pp. 1–18 = *Mélanges Henri Grégoire*, vol. 4 (Brussels: Secrétariat des éditions de l'Institut, 1953); reprinted in Milton V. Anastos, *Studies in Byzantine Intellectual History* (London: Variorum reprints, 1979), article XVII.

globe, it should be 10 feet in diameter, but if one cannot obtain a globe of this size, then one should draw a world map that is 7 feet wide.[64] There is no way to prove that Martellus had Strabo's dictum in mind when designing the map, but the map's dimensions are consistent with such an intention. The map is 201 cm wide; the acanthus border, which was not originally part of the map, is 1.2 cm wide, so the width of the map proper is 198.6 cm, and if this dimension was supposed to be equivalent to 7 feet, that would entail a foot measurement of 28.37 cm. Martellus cannot have been using the Florentine foot, which was a quite long, measuring 32.48 cm, but there are numerous cities in Martellus's native Germany where the foot (as locally defined) was very close to 28.37 cm.[65] Thus even the width of Martellus's map may reflect his study of and respect for the classics.

The map is on a total of 12 sheets of paper of various different sizes: the majority of the map, all of it except the parts in the west and the south, is on six large sheets, each of which measures about 59 × 45 cm; the southern part of the map is on three half-sheets, each about 59 × 21 cm; the western edge of the map is on three smaller strips of paper of differing lengths that measure 29 × 19, 31 × 19, and 48 × 19 cm.[66] These sheets were attached to canvas which has been mounted on a wooden frame, much like a painting. The design of the map was painted on the paper after the sheets were joined.

One of the problems presented by the map is whether it is a purely manuscript map or whether the hand painting was done on top of a printed map of an essentially similar design. Unfortunately, the multispectral images of the map made in 2014 did not provide any evidence as to whether a printed map underlies the painted map, and it was not possible to examine the back of the map at all. But it is worth examining the evidence regarding the possibility that the Yale Martellus map is painted over a printed cartographic substrate.

[64] See Strabo 2.5.10, translated into English in *The Geography of Strabo*, trans. Hans Claude Hamilton and William Falconer (London and New York: G. Bell & Sons, 1903), vol. 1, p. 176: "Any one who is able will certainly do well to obtain such a globe. But it should have a diameter of not less than ten feet: those who cannot obtain a globe of this size, or one nearly as large, had better draw their chart on a plane surface, of not less than seven feet."

[65] For measurements of the foot in various German cities, see Fritz Verdenhalven, *Alte Mess- und Währungssysteme aus dem deutschen Sprachgebiet: was familien- und Lokalgeschichtsforscher suchen* (Neustadt an der Aisch, Germany: Degener, 1993), pp. 19–20; some of the cities whose foot were very close to 28.37 cm are Arnstadt, 28.25 cm; Dresden, 28.333 cm; Erfurt, 23.326 cm; Fulda, 28.288 cm; Gera, 28.319 cm; Leipzig, 28.25 cm; Reuß ältere Linie, 28.32 cm; and Sachsen, 28.311.

[66] R. A. Skelton, "World Map by Henricus Martellus Germanus, c. 1489, at Berne: Supplementary Report," June, 1960 (see note 2 in Front Matter), p. 2, revising his statement that the map was on nine sheets in R. A. Skelton, "World Map by Henricus Martellus Germanus, c. 1489, at Berne," January 10–17, 1960 (see note 2 in Front Matter), p. 2. A diagram of the 12-sheet configuration is attached to the copy of Skelton's original report at the Beinecke, while a diagram of the initially proposed 9-sheet configuration is attached to the copy of Skelton's report that is at the James Ford Bell Library.

During his examinations of the map before its purchase and donation to Yale, Professor Gioia raised a small piece of the paper from the canvas and said that "the design in black had penetrated nearly the whole thickness of the paper. This would not occur in the case of a pen- or brush-drawing, but could only result from forcing of the paper under pressure into the incised and inked lines of an engraved plate."[67] This claim is certainly tantalizing, but iron gall ink from a pen can certainly penetrate paper well, and it is difficult to accept Gioia's claim without being able to verify it.

There is a record of a printed world map that sounds similar to the Yale Martellus map in the inventory of Francsesco Rosselli's map store taken after his son Alessandro's death in 1527.[68] The map in question is listed as "appamondo [*sic*] grande in 9 pezi di 16 fogli chomuni," that is, "large world map on 9 sheets measuring [the equivalent of] 16 *fogli chomuni*." "The *foglio comune* or *foglio mezane* of the fifteenth century measured about 50 × 30 cm, so that the overall dimensions given for Rosselli's large world map would have been 200 × 120 cm, or, allowing for overlaps, about 190 × 110 cm,"[69] which is very close to the dimensions of the Yale Martellus map without the engraved 1.2 cm wide acanthus border that is not part of the map and which was probably added later.[70] However, the fact that the two maps are composed of different numbers of sheets—12 for the Yale Martellus map and nine for the map in Rosselli's inventory—would seem to be an insurmountable difficulty for any attempt to identify the two maps. Moreover, no printer would plan to print the Yale Martellus map on the inconsistently sized 12 sheets that compose the map. The total imprac-

[67] R. A. Skelton, "World Map by Henricus Martellus Germanus, c. 1489, at Berne: Supplementary Report," June, 1960 (see note 2 in Front Matter), p. 1.

[68] This inventory, which is dated February 14, 1527, was published by Jodoco del Badia, "La bottega di Alessandro di Francesco Rosselli merciaio e stampatore (1525)," *Miscellanea fiorentina di erudizione e di storia* 2.14 (1894), pp. 24–30, and subsequently in Giuseppe Boffito and Attilio Mori, *Firenze nelle vedute e piante: studio storico topografico cartografico* (Florence: Tipografia Giuntina, 1926; Rome: Multigrafica, 1973), pp. 146–150; Christian Hülsen, "Die alte Ansicht von Florenz im Kgl. Kupferstichkabinett und ihr Vorbild," *Jahrbuch der Königlich preussischen Kunstsammlungen* 35 (1914), pp. 90–102, esp. 98–102; and Arthur M. Hind, *Early Italian Engraving: A Critical Catalogue with Complete Reproduction of All the Prints Described* (London: B. Quaritch, 1938–48; Nendeln, Liechtenstein: Kraus Reprint, 1970), vol. 1, pp. 304–309. For discussion of the maps listed in the inventory, see Roberto Almagià, "On the Cartographic Work of Francesco Rosselli," *Imago Mundi* 8 (1951), pp. 27–34. Another, possibly earlier version of the inventory has been found in the Archivio di Stato in Florence: see Gentile, *Firenze e la scoperta dell'America* (see note 8 in Chap. 1), pp. 247–250.

[69] R. A. Skelton, "World Map by Henricus Martellus Germanus, c. 1489, at Berne," January 10–17, 1960 (see note 2 in Front Matter), p. 8, citing Konrad Haebler, *Handbuch der Inkunabelkunde* (Leipzig: Verlag Karl W. Hiersemann, 1925), p. 39.

[70] On the acanthus border of the Yale Martellus map, see R. A. Skelton, "World Map by Henricus Martellus Germanus, c. 1489, at Berne," January 10–17, 1960 (see note 2 in Front Matter), pp. 2 and 3. Skelton suggests that the border of the Martellus map is quite similar to that on a late sixteenth-century-printed Ptolemaic world map that may be a reprint of an earlier map, citing George H. Beans, "Some Notes from the Tall Tree Library," *Imago Mundi* 14 (1959), pp. 112–113, where the map is illustrated opposite p. 113, but unfortunately this does not seem to me to shed much light on the question of the source of the border.

ticality of that scheme is particularly apparent when one realizes that one might quite reasonably have printed the Yale Martellus map on eight of the 59 × 46.5 cm sheets on which the Yale map was painted, which would have resulted in a map measuring 118 × 186 cm, close to the 119.6 × 198.6 cm dimensions of the map without the acanthus border. Thus while it seems that the Yale Martellus map was not printed, the existence in Rosselli's inventory of a world map on nine sheets and of dimensions similar to those of the Yale Martellus, together with the known connection between Martellus and Rosselli, leave open the possibility that Rosselli printed something similar to the Yale Martellus map. Various pieces of evidence presented below indicate that Waldseemüller was using a map very similar to the Yale Martellus map, but with some differences, and this hypothetical printed version of the Yale Martellus map is a very strong candidate for the map that Waldseemüller was using.

Martin Behaim used a printed world map that would seem to have been large in the making of his terrestrial globe of 1492, and given that his globe shows the influence of the geography of the Yale Martellus map,[71] it is tempting to think that this printed map may have been closely related to the Yale Martellus map. In August of 1494, Georg Holzschuher submitted to the Nürnberg City Council an account of his expenditures associated with the making and acquisition of Behaim's globe, including the expense for buying from Behaim a printed world map ("getruckte mapa mundy") that Behaim had used as a source for his globe ("die da wol dint zu dem apffel"), and also the expenses for painting, lining, gluing, and framing the map.[72] Arend W. Lang has suggested that this map "doubtlessly was of considerable size,"[73] and it is remarkable that it ended up lined, painted, and framed like the Yale Martellus map.

[71] On the influence of Martellus on Behaim, see the references in note 3 in Front Matter. Campbell, *The Earliest Printed Maps* (see note 54 in Chap. 1), p. 213, notes that the Yale Martellus map is appropriate as a model for Behaim's globe as it is "the only non-Ptolemaic fifteenth-century world map to be graduated, like the globe, with degrees of longitude and latitude," but sees differences between the treatments of southern Africa in Behaim's globe and the Yale Martellus map, and on p. 217 concludes that the Yale Martellus map is purely manuscript—but does not indicate his reasons for this conclusion.

[72] The German text is supplied by Johann Petz, "Urkundliche Beiträge zur Geschichte der Bücherei des Nürnberger Rates, 1429–1538," *Mitteilungen des Vereins für Geschichte der Stadt Nürnberg* 6 (1886), pp. 123–174, at 169: "Item so zalt ich her Merten Beham vmb ein getruckte mapa mundy, da die gantz weit ina wegriffen ist, die da wol dint zu dem apffel vnd in die kantzley gehenckt wirtt, kost 1 fl. 3 lb.; mer kost sie zu maln 1 fl.; mer zu fütern vnd leima 5 lb. 10 dn.; mer dem schreiner in ram vnd zwue taffel gemacht 1 fl.; mer dem Starch maller von disen hulzen tafeln zu maln, 4 lb. 6 dn.; facit...... fl. 4. lb. 4. dn. 6." This text is also supplied by Ernst Georg Ravenstein, *Martin Behaim, His Life and His Globe* (London: G. Philip & Son, Ltd., 1908), pp. 111–112, with an English translation, but Ravenstein mistakenly suggests that Behaim made the map in question. Ravenstein's translation of the whole passage is cited by Edward Luther Stevenson, *Terrestrial and Celestial Globes* (New Haven: Yale University Press, 1921), vol. 1, pp. 56–57. This passage is discussed by R. A. Skelton, "World Map by Henricus Martellus Germanus, c. 1489, at Berne," January 10–17, 1960 (see note 2 in Front Matter), p. 11.

[73] See A. Wilhelm Lang, "Traces of Lost North European Sea Charts of the fifteenth Century," *Imago Mundi* 12 (1955), pp. 31–44, at 41.

Indeed, it has been suggested that the Yale Martellus map must be the result of painting over a printed substrate because the existence of a printed version of the map is the only way to explain its wide influence (on Columbus, Behaim, and Waldseemüller, among others).[74] To my mind, in and of itself, this point carries little force, as it is certainly possible that Martellus produced more than one manuscript map similar to the Yale Martellus map. However, when this point is considered together with the evidence that Rosselli had a large printed world map in his shop, and the evidence that Behaim relied on a large printed map as a source for his 1492 globe, while the globe shows the influence of the geography of the Yale Martellus map, it seems likely that Martellus did create a large printed world map with geography very similar to that of his map at Yale.

An important element of the Yale Martellus map that has received little attention is the printed acanthus border, which, as mentioned above, is not integral to the map but is a later addition.[75] In particular, it has not been remarked that the initials "I H" appear in the acanthus border, specifically in the left border opposite the westernmost point of Africa, in the left border just opposite the top edge of the cartouche above the large text block, in the right border just opposite the top of the large text block, and in the right border opposite the top of the wind-head labeled *Subsolanus* (Fig. 1.7). These initials "I H" in the acanthus border decoration recall the "I H" that appears in the border decorations in the 1522 Strasbourg edition of Ptolemy's *Geography* and also in the later editions that use some of the plates from the 1522 edition, namely, the 1525 (Strasbourg), 1535 (Lyon), and 1541 (Vienna) editions of the *Geography* (Fig. 1.8).[76] It is difficult to imagine that the two creators of border decorations for cartographic contexts with the same initials are not one in the same. Some candidate identifications for I. H. have been proposed,[77] but none of them is certain.

[74] See Carlos Sanz, "Un mapa del mundo verdaderamente importante en la famosa Universidad de Yale," *Boletín de la Real Sociedad Geográfica* 102 (1966), pp. 7–46, esp. 15–19.

[75] R. A. Skelton, "World Map by Henricus Martellus Germanus, c. 1489, at Berne," January 10–17, 1960 (see note 2 in Front Matter), p. 3, suggests that the acanthus border is very similar to that in a printed Ptolemaic world map probably designed in the late fifteenth century that survives in two exemplars: Chicago, Newberry Library, Novacco 4F 3 and Providence, Rhode Island, John Carter Brown Library, Case 12b A480 1. For discussion of this map, see Tony Campbell, *The Earliest Printed Maps, 1472–1500* (London: British Library, 1987), pp. 27–30 with plate 37; and Chet Van Duzer, "Ptolemy from Manuscript to Print: New York Public Library's Codex Ebnerianus (MS MA 97)," *Imago Mundi* 67.1 (2015), pp. 1–11. I do not find the similarity between the acanthus borders of the Yale Martellus map and those of this Ptolemaic world map particularly compelling.

[76] These editions of Ptolemy's *Geography* use maps made by Laurent Fries which were based on maps in Waldseemüller's 1513 edition of Ptolemy. For descriptions of these editions of Ptolemy, see Wilberforce Eames, *A List of Editions of Ptolemy's Geography 1475–1730* (New York, 1886) (reprinted from Joseph Sabin's *Bibliotheca Americana*), pp. 15–17, 17–18, 18–19, and 20–21; and Carlos Sanz, *La Geographia de Ptolomeo, ampliada con los primeros mapas impresos de América, desde 1507* (Madrid: Librería General V. Suárez, 1959), pp. 150–155, 156–164, 169–179, and 187–188. In the 1522 edition, the initials "I H" appear in the decorations in the decoration at the beginning of the text for the eighth map of Europe, the second map of Africa, the second map of Asia, the eighth map of Asia, the twelfth map of Asia, the modern map of England, and the modern map of southern Africa.

[77] Charles Schmidt, *Répertoire bibliographique Strasbourgeois jusque vers 1530* (Baden-Baden: Heitz, 1963), p. 80, suggests that I. H. may be Jean (Hans) Herbst or Herbster, but Hans Koegler in Ulrich

Fig. 1.7 Detail of the initials "I H" in the printed acanthus border of the Yale Martellus map. (Courtesy of the Beinecke Rare Book and Manuscript Library)

Fig. 1.8 Details of the initials "I H" in the border decorations in the 1522 Strasbourg edition of Ptolemy's Geography. *Library of Congress, Rare Book and Special Collections Division, Thacher Collection A823. (Courtesy of the Library of Congress)*

The Yale Martellus map is a much more ambitious production than the world maps in the manuscripts of Martellus's *Insularium* and also than the

Thieme and Felix Becker, eds., *Allgemeines Lexikon der bildenden Künstler von der Antike bis zur Gegenwart* (Leipzig: W. Engelmann, 1907–1950), vol. 16, p. 451, rejects that suggestion, and François Ritter, *Histoire de l'imprimerie alsacienne aux XVe et XVIe siècles* (Strasbourg: Le Roux, 1955), p. 98, mentions Jerome Hopferd of Augsberg as a possibility. There is some discussion of the initials in Hildegard Binder Johnson, *Carta marina: World Geography in Strassburg, 1525* (Minneapolis: University of Minnesota Press, 1963), pp. 44 and 133.

Martellus-Rosselli map— ambitious not only for its size but also for the larger geographical area it covers, its cartographic sophistication, and level of detail. According to the scale of longitude along its southern edge, the Yale Martellus map covers 280° of longitude[78]; the world maps in Martellus's *Insularium* and the Martellus-Rosselli do not have a scale of longitude, but we can estimate the longitudinal width of these other maps by comparing them with the Yale Martellus map. I will make the comparison with the Martellus-Rosselli map. The western edges of the two maps are at essentially the same longitude, but the Yale Martellus map comprises much more of the Indian Ocean, including Japan. The island of Java is almost at the eastern edge of the Martellus-Rosselli map, and if we proceed south from off the eastern coast of Java on the Yale Martellus map along a contour parallel to the curved eastern edge of the map, we strike the scale of longitude at 230° or 40° west of the marker for 270° at the eastern edge of the map. Thus while the Yale Martellus map covers 280° of longitude, the Martellus-Rosselli map covers about 240° of longitude.[79]

The Yale Martellus map thus includes a substantially larger part of the earth's surface than the cartographer's earlier maps and may also be seen as presenting a different conception of the world. Martellus's earlier maps are dominated by land—the boundaries of the maps are drawn close to the known limits of the land—but the Yale Martellus map shows a greater interest in the ocean, showing more of the southern as well as of the eastern Indian Ocean. In addition, the scale of longitude indicates that Martellus clearly understood how much of the world's surface still awaited discovery. Indeed, although there were terrestrial globes before Behaim's of 1492 and also before Martellus's map,[80] it is worth emphasizing that the Yale Martellus map is the only surviving non-Ptolemaic map made prior to Behaim's globe that includes scales of latitude and longitude.

The Yale Martellus map is constructed on a pseudo-cordiform or heart-shaped projection, a modified form of Ptolemy's second projection.[81] The map is the first known use of this projection, and it was no doubt adopted in order to allow the inclusion of more of the eastern Indian Ocean and of Japan. In the border of the map, there are 12 wind-heads with descriptive legends that will be examined below. R. A. Skelton has suggested that the

[78] R. A. Skelton, "World Map by Henricus Martellus Germanus, c. 1489, at Berne," January 10–17, 1960 (see note 2 in Front Matter), p. 4, indicates that it coves 270° of latitude, and the easternmost marker in the graduation of latitude along the southern edge of the map is "270," but there are 10° of latitude beyond 0° (marked "360" and "355") at the western edge of the map.

[79] Almagià, "I mappamondi di Enrico Martello" (see note 4 in Front Matter), p. 306, estimates the width of the Martellus-Rosselli map at 220°, but he did not have the Yale Martellus map to use for comparison, as it was not discovered until after his article was published in 1940.

[80] See Józef Babicz, "The Celestial and Terrestrial Globes of the Vatican Library, Dating from 1477, and their Maker Donnus Nicolaus Germanus (ca 1420–ca 1490)," *Der Globusfreund* 35–37 (1987), pp. 155–168; and Patrick Gautier Dalché, "Avant Behaim: les globes terrestres au XVe siècle," *Médiévales* 58 (2010), pp. 43–61.

[81] For references on the projection that Martellus used, see note 15 in Front Matter.

artistic style of these wind-heads recalls that used in the wind-heads on Berlinghieri's world map in his *Geographia* of 1482, the Martellus-Rosselli map, and Giovanni Matteo Contarini's world map of 1506,[82] but I do not find these resemblances to be particularly close. It is worth remarking that the wind-heads on the Yale Martellus map are not particularly similar to those on the world map in either of the two manuscripts of Ptolemy's *Geography* made by Martellus, namely, Vatican City, Biblioteca Apostolica Vaticana, Vat. lat. 7289, ff. 1v–2r, and Florence, Biblioteca Nazionale Centrale, Magliabechiano XIII 16, ff. 88v–89r. It seems likely, then, that Martellus hired a different artist to paint the wind-heads on each of these maps.

An interesting characteristic of the Yale Martellus map that I have not seen remarked upon is that the layout is asymmetrical, in that the map is tilted about 2.4° to the left with respect to the rectangular borders of the map (see Fig. 1): the wind-head of Auster or Notus blows at the midpoint of the scale of longitude (at the center of the bottom edge of the map proper), but the outline of the map proper curves higher on the right or eastern side than it does on the left or western side, as may be appreciated by comparing the height of the southwestern and southeastern "corners" of the map, the heights of the points where the tropics and equator reach the western and eastern edges of the map (and the wind-heads located at the western and eastern ends of the equator), and the paucity of space for the wind-head north of the eastern end of the Tropic of Cancer compared with the abundant space for the wind-head north of the western end of the same Tropic.

The fact that this tilting of the Yale Martellus map is not accidental is confirmed by the presence of similar tilting, though somewhat less pronounced, on the Martellus-Rosselli map (see Fig. 1.6): the southeastern "corner" of the map proper is higher than the southwestern; the equator touches the eastern border of the map higher than the western; and the northeastern inflection point in the map's border is higher and sharper than the western. John Hessler suggested to me that this feature represents a naïve attempt to correct for magnetic declination, that is, the fact that magnetic north and true north do not coincide; this attempt would be naïve in that magnetic declination varies from place to place. Joaquim Alves Gaspar pointed out that if the intention of the tilting was to compensate for magnetic declination in the eastern Atlantic, then the compensation was made in the wrong direction: the declination in the eastern

[82]The unique surviving exemplar of Contarini's map is in the British Library, Maps C.2.cc.4. It is well reproduced in Nebenzahl, *Atlas of Columbus* (see note 42 in Chap. 1), pp. 45–47, and there is a good brief discussion of the map with an illustration in Shirley, *The Mapping of the World* (see note 54 in Chap. 1), pp. 23–25; there is also a facsimile published as Giovanni Matteo Contarini, *A Map of the World, Designed by Gio. Matteo Contarini, engraved by Fran. Roselli 1506* (London: Printed by order of the Trustees, sold at the British Museum, 1924); for discussion of the map, see Edward Heawood, "A Hitherto Unknown World Map of A. D. 1506," *Geographical Journal* 62.4 (1923), pp. 279–293.

Atlantic was positive around 1500,[83] but the tilting of the map would correct for a negative declination. Nonetheless, it is difficult to imagine another motivation for this tilting, if not to try to compensate for magnetic declination. Waldseemüller's 1507 map, incidentally, shows no similar tilting (Fig. 2).

With regard to the geographical content of the map, it represents the world from 90°N to 40°S and from 10° west of the Ptolemaic prime meridian to 270°E, including the island of Japan at the eastern edge of the map. Japan had been represented on two earlier European maps, namely, Albertin de Virga's world map of c. 1411–1415, where there is a large island in the southeast labeled *Caparu sive Java magna*, so that the cartographer identifies two islands which are quite distinct in Marco Polo[84] and on Fra Mauro's *mappamundi* of c. 1450, where it is labeled *Ixola de cimpagu*.[85] Martellus had shown interest in Japan earlier when he made a separate map of it in the Florence manuscript of his *Insularium* (Biblioteca Medicea Laurenziana, Pluteo 29.25, f. 76r) (Fig. 1.9), but the island does not appear in any of his other world maps.[86]

Martellus's representation of Europe is largely Ptolemaic, and this is a puzzling choice, at least at first glance: in the world maps in his *Insularium*, he gives Spain, the British Isles, and the Black Sea more modern shapes derived from nautical charts. His mistaken depiction of Greenland as a long peninsula jutting to the southwest, which also appears in his other world maps, is based on Ptolemaic maps by Donnus Nicolaus Germanus (and proved very influential). As it seems that the Yale map was made after Martellus's other maps (see below), his choice to use the older Ptolemaic cartographic model for Europe in the Yale map is surprising.[87]

Eastern Asia on the Yale map is dense with legends, many of which come from Marco Polo. On the northern coast of Asia, near the center of the map, there is a distinctive peninsula that juts westward in the northern

[83] See Ricardo Cerezo Martínez, *La cartografía náutica española en los siglos XIV, XV y XVI* (Madrid: C.S.I.C., 1994), p. 105.

[84] For discussion and a reproduction of the Albertin de Virga world map, see Destombes, *Mappemondes* (see note 10 in Front Matter), pp. 205–207 and plate 28; the current location of the map is unknown, but it was reproduced in facsimile in Franz von Wieser, *Die Weltkarte des Albertin de Virga ausdem Anfange des XV. Jahrhunderts in der Sammlung Figdom in Wien* (Innsbruck: Heinrich Schurick, 1912), and in Youssouf Kamal, *Monumenta cartographica Africae et Aegypti* (Cairo, 1926–1951), vol. 4, fasc. 3, f. 1377.

[85] Piero Falchetta, *Fra Mauro's World Map*, trans. Jeremy Scott (Turnhout: Brepols, and Venice: Biblioteca Nazionale Marciana, 2006), pp. 434–435, *1334.

[86] The map of Japan in the Florence manuscript is illustrated in Roberto Almagià, "I mappamondi di Enrico Martello e alcuni concetti geografici di Cristoforo Columbo," *La Bibliofilia* 42 (1940), pp. 288–311, esp. 303, and Gentile, *Firenze e la scoperta dell'America* (see note 8 in Chap. 1), plate 46. For discussion see George Kish, "Two Fifteenth-Century Maps of 'Zipangu': Notes on the Early Cartography of Japan," *The Yale University Library Gazette* 40.4 (1966), pp. 206–214.

[87] For discussion of the evolution of Waldseemüller's preference for nautical charts over Ptolemy as a cartographic model, see my introduction to Waldseemüller's *Carta marina* in John W. Hessler and Chet Van Duzer, *Seeing the World Anew: The Radical Vision of Martin Waldseemüller's 1507 & 1516 World Maps* (Washington, DC: Library of Congress, and Delray Beach, FL: Levenger Press, 2012), pp. 49–68.

Fig. 1.9 The map of Japan in the Florence manuscript of Martellus's island book (Biblioteca Medicea Laurenziana, Pluteo 29.25, f. 76r). By concession of the Ministero dei Beni e delle Attività Culturali e del Turismo. (Any further reproduction by any means is prohibited)

ocean, which appears on the world maps in Martellus's *Insularium* and also on the Martellus-Rosselli map. In eastern Asia, what on Martellus's other world maps is a small southward-jutting peninsula becomes a huge triangular westward-jutting peninsula whose southern edge is aligned with the Tropic of Cancer. Obviously the greater longitudinal range of the Yale Martellus map allowed more room for this peninsula, but there is no way to know whether on the maps in manuscripts of the *Insularium*, the small peninsula was intended to represent a larger peninsula for which there was insufficient room or whether Martellus changed his ideas about the geography of eastern Asia and came to believe that the peninsula must be much larger. In Southeast Asia there is a large peninsula stretching toward the southwest; this peninsula,[88] which is in part a remnant of Ptolemy's mythical land bridge that joined southern Africa and southern Asia,[89] also appears on the world maps in Martellus's *Insularium*. Both the large triangular eastward-jutting peninsula and the large peninsula in Southeast Asia appear on two important maps influenced by Martellus, Martin Behaim's globe of 1492 and Martin Waldseemüller's 1507 world map.

In the eastern Indian Ocean, in addition to Japan in the north, the Yale Martellus map has the islands of Seylam, Peutam, Necura, Angama, and Java Major, all of which come from Marco Polo (these same islands appear with names on the Martellus-Rosselli map, but not on Martellus's other world maps). In the western Indian Ocean, the depiction of Taprobana is Ptolemaic, while Madagascar, which is located just off the coast of south-eastern Africa, is from Marco Polo. It is surprising that Zanzibar is absent from the Yale Martellus map, since it appears on the Martellus-Rosselli map (see Fig. 1.6), and there is plenty of room for it on the Yale Martellus map, which in fact extends further to the south than the Martellus-Rosselli map. Zanzibar does appear on Waldseemüller's 1507 map. This is one of several pieces of evidence that suggest that Waldseemüller was using a different version of Martellus's 280° image of the world than we have in the Yale map.

Given Martellus's heavy use of Marco Polo, the question arises as to which version of the text he was using. The first printed edition of Marco Polo is a German translation published in 1477,[90] but as Martellus supplies his legends from Marco Polo in Latin, it is tempting to think that he made use of the earliest printed Latin edition, which is of the Latin translation by Francesco Pipino and was published between 1483 and 1485[91]: no other

---

[88] On the "Tiger-Leg" or "Cattigara" peninsula, see note 16 in Front Matter.

[89] On the Ptolemaic land bridge joining Africa and Asia, see Ptolemy *Geography* 7.5, translated in A. Jones and J. L. Berggren, eds., *Ptolemy's Geography: An Annotated Translation of the Theoretical Chapters* (Princeton: Princeton University Press, 2000), pp. 108–110; and Wilcomb E. Washburn, "A Proposed Explanation of the Closed Indian Ocean on Some Ptolemaic Maps of the 12th–15th Centuries," *Revista da Universidade de Coimbra* 32 (1986), pp. 431–441.

[90] The *editio princeps* of Marco Polo is *Hie hebt sich an das puch des edel[e]n Ritters vn[d] landtfarers Marcho Polo* (Nuremberg: Friedrich Creussner, 1477); for discussion of this edition, see Marco Polo, *The Book of Ser Marco Polo, the Venetian: Concerning the Kingdoms and Marvels of the East*, transl. and ed. Henry Yule (3rd edn., London: J. Murray, 1903), vol. 2, pp. 554–556.

[91] The first Latin edition of Marco Polo is *De consuetudinibus et condicionibus Orientalium regionum* (Gouda: Gerard Leeu, c. 1483–1485). This edition has been reproduced in facsimile twice, first as *Itinerarium, Antverpiae, 1485* (Tokyo: [s.n.], 1949); the second facsimile is of the copy annotated by Christopher Columbus, *Libro de las maravillas del mundo: facsímil del que, usado por Cristóbal Colón, se encuentra depositado en la Biblioteca Capitular y Colombina del Cabildo Catedral de Sevilla* (Madrid: Testimonio, 1986), with two volumes of transcription, translation, and commentary by Juan Gil under the titles *El libro de Marco Polo* and *The Book of Marco Polo*. For discussion of this edition, see Yule, *The Book of Ser Marco Polo* (see note 90 in Chap. 1), vol. 2, pp. 558–559; Lotte Hellinga, "Marco Polo's Description of the Far East and the Edition Printed by Gheraert Leeu," in Elly Cockx-Indestege, ed., *E codicibus impressisque: opstellen over het boek in de Lage Landen voor*

Latin edition was published until 1532. But as will become clear from the phrasing of several of Martellus's legends transcribed below, he was not using the printed edition, but rather a manuscript. I have not identified the specific manuscript that he was using—and indeed, since the vast majority of medieval manuscripts do not survive,[92] it is very likely that the manuscript is no longer extant—but the manuscript that Martellus used was similar to a mid-fifteenth-century manuscript made in Italy which is now in Naples, Biblioteca Nazionale Vittorio Emanuele III, Vind. lat. 50.[93] I will therefore be citing the text of this manuscript for comparison with the legends on Martellus's map.

The Yale map's depiction of Africa is remarkable. The shape of the northern part of the continent, with its sharp northwestern corner, is Ptolemaic, despite Martellus's knowledge of recent Portuguese discoveries, as witnessed by the legends on the world map in the London manuscript of his *Insularium* and on the Martellus-Rosselli map and by his depiction of the northern part of the continent with a more modern shape in the world maps in his *Insularium* and in the Martellus-Rosselli map. Indeed, a legend on the world map in the London manuscript of the work, discussed above, explicitly indicates his pride in displaying the correct and modern shape of the continent: "This is the modern true shape of Africa according to the description of the Portuguese, from the Mediterranean Sea to the southern ocean." There is also a legend about the Portuguese discoveries on the Yale Martellus map, at the southern tip of Africa. It reads[94]:

> *huc usque columpnam et ilha de fonti sive penedo pervenit classis Johan[nis] secundi*
> *To this point, to the column and the Isle (or Rock) of the Fountain, came the fleet of John the Second.*

Elly Cockx-Indestege (Louvain: Peeters, 2004), vol. 1, pp. 309–328, revised version published as "The Travels of Marco Polo and Gheraert Leeu," in *Texts in Transit: Manuscript to Proof and Print in the Fifteenth Century* (Leiden and Boston: Brill, 2014), pp. 278–303.

[92] On the survival rates of medieval manuscripts, see John L. Cisne, "How Science Survived: Medieval Manuscripts' 'Demography' and Classic Texts' Extinction," *Science* 307.5713 (February 25, 2005), pp. 1305–1307, with criticism in Georges Declercq, "Comment on 'How Science Survived: Medieval Manuscripts' 'Demography' and Classic Texts' Extinction," *Science* 310.5754 (December 9, 2005), p. 1618b; and Eltjo Buringh, "Losses of Medieval Manuscripts," in his *Medieval Manuscript Production in the Latin West: Explorations with a Global Database* (Leiden and Boston: Brill, 2011), pp. 179–251, esp. 231–232. Cisne indicates a manuscript survival rate of one in seven from some ninth-century Carolingian workshops, but this number is implausibly high, and Buringh's figure of one in 16.6 for fifteenth-century manuscripts seems much better founded.

[93] For a description of the Naples manuscript, see Consuelo Wager Dutschke, "Francesco Pipino and the Manuscripts of Marco Polo's Travels," Ph.D. Dissertation, University of California at Los Angeles, 1993, pp. 788–794. The text of this manuscript is edited in Marco Polo, *Marka Pavlova z Benátek, Milion: Dle jediného rukopisu spolu s prilusnym zakladem latinskym*, ed. Justin Václav Prásek (Prague: Nákl. Ceské akademie císare Frantiska Iozefa, 1902).

[94] This legend is legible in natural light, and is mentioned by Roberto Almagià, "Worldmap by Henricus Martellus at Berne," June 1960 (see note 2 in Front Matter), p. 2; R. A. Skelton, "World Map by Henricus Martellus Germanus, c. 1489, at Berne: Supplementary Report," June 1960 (see note 2 in Front Matter), p. 3; Destombes, *Mappemondes* (see note 10 in Front Matter), p. 230; copied by Sanz, "Un mapa del mundo verdaderamente importante" (see note 3 in Front Matter), p. 10.

This legend refers to the voyage of Bartolomeu Dias in 1487–1488 and the *padrão* or commemorative column that he set up at the furthest point he reached and is very similar to a legend on the Martellus-Rosselli map examined above.

Martellus's choice in his map at Yale to follow Ptolemy in his depiction of northern Africa rather than models he believed to be more accurate recalls his similar choice in depicting Europe. This choice, which seems strange today, was reasonable in the context of late fifteenth- and early sixteenth-century cartography. Martin Waldseemüller in his 1507 world map also chose to depict northern Africa with a Ptolemaic shape, though he used as a source at least one map, the manuscript world map made c. 1504 by Nicolo de Caverio,[95] which depicts Africa according to the contours that had been determined during more recent voyages, rather than according to Ptolemy (see Fig. 1.10). Moreover, the habit of adding *Tabulae modernae*, or modern maps, to both manuscripts and printed editions of Ptolemy is strong evidence that both ways of seeing the world (Ptolemaic and modern) were thought to have validity well into the sixteenth century.[96]

The contours of southern Africa on the Yale Martellus map are unusual and in fact unique among surviving fifteenth- and sixteenth-century maps. Martellus depicts Africa as circumnavigable, rather than with a Ptolemaic land bridge joining southern Africa and southern Asia; this is not at all surprising, as he did so in all of the surviving world maps in his *Insularium* and in the Martellus-Rosselli map.[97] But the shape of southern Africa on the Yale map is very different from what we see in the world maps in his *Insularium*, the Martellus-Rosselli map, or indeed in any other

[95] The Caverio chart is in Paris, Bibliothèque nationale de France, Cartes et plans, SH archives 1. The chart is well reproduced in Kenneth Nebenzahl, *Atlas of Columbus* (see note 42 in Chap. 1), pp. 41–43 and at a larger scale in the 11-sheet black-and-white facsimile that accompanies Edward L. Stevenson's study of the map, *Marine World Chart of Nicolo de Caneiro Januensis 1502 (circa)* (New York: American Geographical Society and the Hispanic Society of America, 1908). There is also a color facsimile: *Planisphère nautique sur vélin du Génois Nicolao de Caverio* (Paris: Bibliothèque nationale, 1992) but on a much smaller scale than the original. For discussion of Waldseemüller's use of the Caverio map as a source, see Fischer and von Wieser, *Die älteste Karte mit dem Namen Amerika* (see note 14 in Front Matter), pp. 26–29.

[96] For discussion of the history of adding *Tabulae modernae* to manuscripts and printed editions of Ptolemy, see the references in note 10 in Chap. 1.

[97] Earlier maps that show Africa to be circumnavigable include Albertin de Virga's world map of c. 1411–1415 (see note 84 in Chap. 1), the world map in the so-called Medici Atlas (Florence, Biblioteca Medicea Laurenziana, MS Gaddi 9, ff. 2v-3r), which was perhaps first made c. 1351 but probably retouched in the fifteenth century: for a reproduction see Theobald Fischer, *Raccolta di mappamondi e carte nautiche del XIII al XVI secolo* (Venice: F. Ongania, 1881), and for discussion see George H. T. Kimble, "The Laurentian World Map with Special Reference to Its Portrayal of Africa," *Imago Mundi* 1 (1935), pp. 29–33, with a fuller discussion in G. H. Kimble, "The Mapping of West Africa in the Fourteenth and Fifteenth Centuries as Illustrative of the Development of Geographical Ideas," MA thesis, University of London, 1931, summarized in *Bulletin of the Institute of Historical Research* 9 (1931–1932), p. 190–193; and more recently Relaño, *The Shaping of Africa* (see note 43 in Chap. 1), p. 124. The other important map that shows Africa to be circumnavigable before Martellus is Fra Mauro's world map of c. 1460, and Fra Mauro discusses this issue in one of his legends: see Falchetta, *Fra Mauro's World Map* (see note 85 in Chap. 1), pp. 192–192, *53.

Fig. 1.10 World map by Nicolo de Caverio, c. 1504 (Paris, Bibliothèque nationale de France, Cartes et plans, SH archives 1). By permission of the Bibliothèque nationale de France

map by any cartographer. Skelton aptly compares the shape of southern Africa on the Yale Martellus map to a foot, with the heel to the west and the toe to the east.[98] In his earlier maps, Martellus betrays considerable uncertainty with regard to the configuration of southeastern Africa. On the world map in the Florence manuscript of his *Insularium* (Fig. 1.2), which is probably the earliest, he shows a large peninsula jutting eastward from the southern end of the continent, and the peninsula was larger before he revised it. On the map in the Leiden manuscript (Fig. 1.3), as mentioned above, he does not draw the coastline of this part of the continent. On the map in the London manuscript (Fig. 1.4), the peninsula has returned, but Martellus avoids committing as to its full eastward extension by placing not one but two scrolls just to the east of it, one about the Portuguese discoveries and the other reading *Oceanvs Indicvs Meridional*[*is*]. On the Martellus-Rosselli map (Fig. 1.6), there is no eastward-jutting peninsula in southern Africa, a remarkable change from his other maps.

None of his other maps shows anything like the westward protuberance (the "heel" of the foot) in southern Africa on the Yale Martellus map, and there are remarkable differences between the interiors of southern Africa on the Yale map and Martellus's other world maps. On his other maps, the interior of the continent south of the Mountains of the Moon is largely empty: the coastal toponyms take up some of the space in the interior, and on the London map (Fig. 1.4), some mountains parallel the coast in the southwestern part of the continent, but in essence the interior is vacant. The situation is entirely different on the Yale Martellus map (and much easier to appreciate in the multispectral images of the map—see Fig. 1.11): there is a complex river system with multiple named rivers south of the Mountains of the Moon, and this system extends almost to the southeastern tip of the continent; moreover there are mountains, lakes, and most significantly named cities in an area that is blank on Martellus's other maps. Clearly Martellus was using a different source for this region in his Yale map than he had used in his other world maps. I will explore this matter in detail below.

I will now examine the question of the date of the Yale Martellus map. Like Martellus's other world maps, it was certainly made after 1489, when report of Bartolomeu Dias's voyage—which departed around August of 1487 and returned in December of 1488—would have been diffused, and Martellus must have developed the geography depicted on the map before 1492, when Martin Behaim made his globe and borrowed the large triangular eastward-jutting Asian peninsula from Martellus: this peninsula does not appear on Martellus's other surviving maps, only on the Yale map. That is, the map at Yale may have been made after 1492, but the geography it depicts was worked out before 1492. The fact that Martellus seems

[98] R. A. Skelton, "World Map by Henricus Martellus Germanus, c. 1489, at Berne," January 10–17, 1960 (see note 2 in Front Matter), p. 6

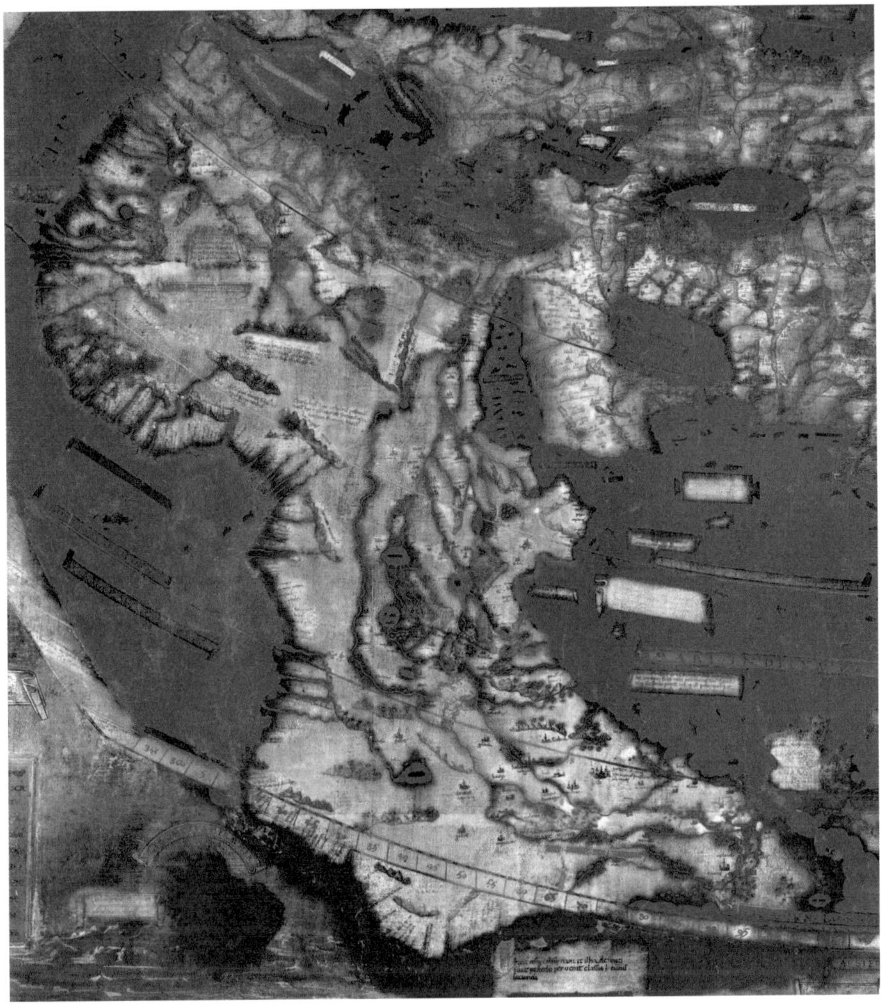

Fig. 1.11 Multispectral image of Africa on the Yale Martellus map. (Image by Lazarus Project/ MegaVision/RIT/EMEL, courtesy of the Beinecke Rare Book and Manuscript Library)

to have made the manuscripts of his *Insularium* in this same general period (though some of them may have been made later than 1492) suggests that 1489 to the mid-1490s was a time of intense cartographic activity for Martellus.

It is possible to be more precise about the dating of the Yale map. The way that Africa crosses the southern edge of the map is different than on Martellus's other maps: on the Yale map, although the continent crosses the edge, that edge, which is formed by the scale of longitude, remains intact. But the way that the water surrounding the land appears outside the edge of the map proper (which does not happen on the Leiden map or the Martellus-Rosselli map) recalls the London map and suggests that the Yale map, like the London map, is later rather than earlier among Martellus's surviving maps. Moreover, the fact that the interior of southern Africa indicates the use of a source that he did not use in the *Insularium* maps also suggests that the Yale map is later in the sequence rather than earlier. This same point could be taken as indicating that the Yale map is earlier than the others, and that Martellus used this different source in that map, but then rejected it before making the others, but given the other evidence that suggests that the Yale map is later, it seems more reasonable to take this difference in the depiction of southern Africa as indicating that the Yale map is later.

Further, in several of the legends on the Yale map, Martellus made use of the *Hortus Sanitatis*,[99] an illustrated encyclopedia that was first published in 1491.[100] For example, in the Indian Ocean northeast of Madagascar, there is a banner bearing this legend:

> *Hic nascitur conchila maris qui murex dicitur que circumiso ferro lacrimas rubeas emittit ex qua fit purpura que ostium appellant.*
>
> *Here is born the sea-shell called the murex, which, when it is cut with iron, sheds red tears from which is made the purple dye called* ostrum.

The murex is mentioned by Pliny in his *Naturalis historia* (9.60.125–126), but his description is not similar to Martellus's legend. The descriptions in Isidore's *Etymologiae* (12.6.50),[101] the second- and third-family

[99] I first noted Martellus's use of the *Hortus Sanitatis* in my lecture "Evidence for a Lost Map Used by Waldseemüller in his Depiction of Eastern Africa and the Indian Ocean," delivered May 15, 2009, at the conference "Exploring Waldseemüller's World," May 14–15, 2009, at the Library of Congress, Washington, DC. A video of that talk is available at http://www.loc.gov/today/cyberlc/feature_wdesc.php?rec=4569, where it is the third talk in the panel.

[100] The *Hortus Sanitatis* "major," which is the work that interests us here, is to be distinguished from the *Hortus Sanitatis* "minor," which is a Latin translation of the German herbal often titled *Gart der Gesundheit*, first published by P. Schoeffer, Mainz, 1485. The herbal published in 1485 has 435 chapters, while the *Hortus Sanitatis* "major" of 1491 has 1066 chapters. There is some very brief discussion of the different editions of each work in J. Christian Bay, "Hortus sanitatis," *Papers of the Bibliographical Society of America* 11.2 (1917), pp. 57–60, but his paper is in essence a call for further research; details about and discussion of the early editions of the *Hortus Sanitatis* are provided by Arnold C. Klebs, "Herbals of fifteenth Century," *Papers of the Bibliographical Society of America* 11 (1917), pp. 75–92, and 12 (1918), pp. 41–57, esp. pp. 48–51 and 54–57. There is a more detailed discussion in Joseph Frank Payne, "On the 'Herbarius' and 'Hortus sanitatis'," *Transactions of the Bibliographical Society* 6.1 (1901), pp. 63–126, esp. pp. 105–124. The first edition of the work was published in Mainz by Jacob Meydenbach, June 23, 1491. For a recent discussion of the book, but with emphasis on the section on plants, see Brigitte Baumann and Helmut Baumann, *Die Mainzer Kräuterbuch-Inkunabeln 'Herbarius Moguntinus' (1484), 'Gart der Gesundheit' (1485), 'Hortus sanitatis' (1491): wissenschaftshistorische Untersuchung der drei Prototypen botanisch-medizinischer Literatur des Spätmittelalters* (Stuttgart: Hiersemann, 2010), pp. 177–222. PDFs of copies of the *Hortus sanitatis* published in 1491 are available from several libraries, including the US National Library of Medicine in Bethesda, and the Bibliothèque Municipale in Montpelier. The part of the *Hortus Sanitatis* devoted to sea creatures has recently been edited and translated into French in Catherine Jacquemard, Brigitte Gauvin, and Marie-Agnès Lucas-Avenel, eds., *Hortus sanitatis: Livre IV, Les poisons* (Caen: Presses universitaires de Caen, 2013), and this edition is available on the internet at http://www.unicaen.fr/puc/sources/depiscibus/accueil.

[101] Here is the text from Isidore 12.6.50: *Murex cochlea est maris, dicta ab acumine et asperitate, quae alio nomine conchilium nominatur, propter quod circumcisa ferro lacrimas purpurei coloris emittat, ex quibus purpura tingitur: et inde ostrum appellatum quod haec tinctura ex testae humore elicitor*; and an English translation from Isidore of Seville, *The Etymologies of Isidore of Seville*, trans. Stephen A. Barney, W. J. Lewis, J. A. Beach, and Oliver Berghof (Cambridge, UK, and New York: Cambridge University Press, 2006), p. 262: "The purple-fish (murex) is a shellfish of the sea, named for its sharpness and roughness. It is called by another name, conchilium (also meaning 'a purple dye'), because when it is cut round with a blade, it sheds tears of a purple color, with which things are dyed purple. And from this ostrum (i.e. purple dye) is named, because this dye is drawn out from the liquid of the shell."

bestiaries,[102] Thomas of Cantimpré's *De natura rerum* (7.54)[103] and the *Hortus Sanitatis* ("De piscibus," Chapter 58)[104] are all much closer, and the text in the *Hortus Sanitatis* is the closest to the legend on Martellus's map. Below I will discuss other legends on Martellus's map that confirm that he was using the *Hortus Sanitatis*.

Although the *Hortus Sanitatis* was very popular, was published in a number of editions, and was translated into German, French, and English,[105] no manuscript of the work is known, and its authorship is uncertain. It seems most likely that the book was the enterprise of Jacob Meydenbach, the publisher of the first edition, who probably acted as editor, organizing the efforts of multiple authors, artists, and woodcutters.[106] Thus Martellus could not have had access to the book in manuscript, and it seems unlikely that he would have access to any sort of pre-publication version of the text, so the date of his Yale map must be after June 1491. As it seems likely that this map or one very similar to it was used by Martin Behaim as a source in creating his globe of 1492, we can date the geography of the Yale Martellus map to late 1491 or early 1492, and it is reasonable to place the Yale Martellus map itself in that same range or soon thereafter.

Below I will demonstrate that Waldseemüller made use of either the Yale Martellus map, or (much more likely) a printed map very similar to it, in the creation of his 1507 map. The question arises of how Waldseemüller obtained the map and indeed how he obtained the various maps and books he used in the creation of his 1507 and 1516 world maps. The most likely explanation is the involvement of a patron, such as René II, Duke of Lorraine, who had an interest in cartography: Waldseemüller and his col-

---

[102] See Willene B. Clark, *A Medieval Book of Beasts: The Second-Family Bestiary: Commentary, Art, Text and Translation* (Woodbridge: Boydell, 2006), p. 213; and Ilya Dines, "A Critical Edition of the Bestiaries of the Third Family," Ph.D. Dissertation, Hebrew University of Jerusalem, 2008, Chapter 108; both bestiary families merely quote Isidore here.

[103] Thomas of Cantimpré, *Liber de natura rerum*, ed. H. Boese (Berlin and New York: De Gruyter, 1973), p. 267: *Murices, sicut dicit Plinius, conche marine sunt, que latent contra Canis ortum, exeuntque statuo tempore. Pretiosum etiam liquorem tingendis vestibus utilem habent. Sed muricibus color in sola vena candida reperitur: reliquum corpus sterile est. Vivis hic color exprimitur, quia cum vita seccum evomunt.* There is an English translation of the longer version of this passage in the Granada manuscript of Thomas in Luis García Ballester, ed., *De natura rerum (lib. IV-XII)* (Granada: Universidad de Granada, 1973–1974), vol. 2, p. 304.

[104] Here is the text from the *Hortus Sanitatis*: *Murix. Isido. Murix est cochlea maris ab acumine et asperitate dicta. que alio nomine conchilium dicitur: eo quod circumcisa ferro lachrymas purpurei coloris emittat. ex quibus purpura tingitur. et inde ostrum appellatum est. quod hec tinctura ex humore teste elicitur.*

[105] On later editions and translations of the *Hortus Sanitatis*, see Payne, "On the 'Herbarius' and 'Hortus sanitatis'" (see note 100 in Chap. 1), pp. 115–124; Payne was not aware of the English translation of c. 1527, which is *The noble lyfe a[nd] natures of man of bestes, serpentys, fowles a[nd] fisshes [that] be moste knoweu* [sic] (Antwerp: Emprented by me Ioh[a]n of Doesborowe, 1527?); this edition has been published in facsimile by Noel Hudson, *An Early English Version of Hortus Sanitatis* (London: B. Quaritch, 1954).

[106] So Payne, "On the 'Herbarius' and 'Hortus sanitatis'" (see note 100 in Chap. 1), p. 115; on p. 114 he quotes Meydenbach's epilog and colophon describing the work, in which he claims all of the credit for the book for himself.

league Matthias Ringmann dedicated the account of Vespucci's voyages in the *Cosmographiae introductio* to René, and Waldseemüller presented to René a copy of that book, his globe of 1507 and the 1507 map.[107] Perhaps René owned a Martellus map and made it available to Waldseemüller, or perhaps he had taken an interest in the project of the 1507 world map and supported the project financially. It is also possible that Waldseemüller's colleague Matthias Ringmann[108] acquired Martellus's map on a trip he made to Italy in 1505 in order to pick up from Gianfrancesco Pico della Mirandola (c. 1470–1533) a manuscript of his (Gianfrancesco's) works and take them back to Thomas Wolf.[109] But even if the acquisition of maps was part of Ringmann's goal for the trip, he must have obtained the substantial funds necessary for those purchases somewhere, and so we are led back to René or another patron.

---

[107] Waldseemüller also records René II's enthusiastic reception of his 1507 map and other works in the dedicatory letter in Ringmann's *Instrvctio manvdvctionem prestans in Cartam itinerariam* (Strasbourg: Grüninger, 1511): this passage is quoted and translated into French by M. d'Avezac, *Martin Hylacomylus Waltzemüller, ses ouvrages et ses collaborateurs* (Paris: Challamel âiné, 1867), pp. 136–137, and into English by Toby Lester, *The Fourth Part of the World: The Race to the Ends of the Earth, and the Epic Story of the Map that Gave America its Name* (New York: Free Press, 2009), p. 373.

[108] On Ringmann see Charles Schmidt, "Mathias Ringmann, humaniste alsacien et lorrain," *Mémoires de la Société d'Archéologie Lorraine* 25 (1875) pp. 165–233; and Charles Schmidt, *Histoire littéraire de l'Alsace à la fin du XVe siècle* (Paris: Sandoz and Fischbacher, 1879–1880), vol. 2, pp. 87–131.

[109] Ringmann's 1505 journey to Italy is alluded to in the introductory epistle of the 1513 edition of Ptolemy's *Geography*; for discussion of the journey, see Charles Schmidt, *Histoire littéraire de l'Alsace*, vol. 2, pp. 96–99. For the suggestion that Ringmann brought charts back with him on this trip, see Toby Lester, *The Fourth Part of the World: The Race to the Ends of the Earth, and the Epic Story of the Map that Gave America its Name* (New York: Free Press, 2009), p. 352. Incidentally Ringmann made another trip to Italy in 1508: see Schmidt, *Histoire littéraire*, vol. 2, pp. 117–118 and Karl A. E. Enenkel, "The Making of Sixteenth-Century Mythography: Giraldi's *Syntagma de Musis* (1507–1511 and 1539), *De deis gentium historia* (ca 1500–1548) and Julien de Havrech's *De cognominibus deorum gentilium* (1541)," *Humanistica Lovaniensia* 51 (2002), pp. 9–53, esp. 43–45.

Chapter 2

The Legends on the Yale Martellus Map

Using the 2014 multispectral images,[1] I have been able to read many of the legends on the Yale Martellus map, and I present those legends, together with English translations and commentary, in the following pages. My purpose in studying the legends has been twofold: first, to gain a deeper understanding of the map itself, which is one of the most important and influential of the late fifteenth and early sixteenth centuries and yet is almost entirely unstudied, and, second, to explore the nature of the relationship between the Martellus map and Martin Waldseemüller's famous world map of 1507. As mentioned in my introductory remarks above, the general similarity between the Yale Martellus map and Waldseemüller's 1507 map has been noted previously, but the extent to which Waldseemüller might have used the Martellus map as a source for the details of his map had never been investigated.

When the cartographic sources of Waldseemüller's 1507 map have been discussed in the past, the focus has been on the world map made c. 1504 by Nicolo de Caverio (Fig. 2.1).[2] Waldseemüller took most of his New World toponyms and also most of those on the western coast of Africa from this map—and we can be quite confident that he was using the surviving Caverio chart, rather than a similar map, because he even copied errors on Caverio's chart.[3] But Caverio's chart is very different from Waldseemüller's, particularly with regard to the amount of information it provides about the interior of the continents. While northern Asia and Central Africa on Caverio's chart are almost devoid of legends or depictions of geographical features, these same regions are full of texts and geographical features on Waldseemüller's map. In what follows I will show that Martellus's map was an important source for these elements of Waldseemüller's map, that is, precisely those elements that give Waldseemüller's map some of its encyclopedic character.[4] In fact, I will show that Martellus's map was a more important source for Waldseemüller's 1507 map than Caverio's map.

Electronic supplementary material: The online version of this chapter (https://doi.org/10.1007/978-3-319-76840-3_2) contains supplementary material, which is available to authorized users.

[1] A set of multispectral images of all 55 of the "tiles" that the Yale Martellus map was divided into for imaging are supplied as Supplementary Images 2.1, 2.2, 2.3, 2.4, 2.5, 2.6, 2.7, 2.8, 2.9, 2.10, 2.11, 2.12, 2.13, 2.14, 2.15, 2.16, 2.17, 2.18, 2.19, 2.20, 2.21, 2.22, 2.23, 2.24, 2.25, 2.26, 2.27, 2.28, 2.29, 2.30, 2.31, 2.32, 2.33, 2.34, 2.35, 2.36, 2.37, 2.38, 2.39, 2.40, 2.41, 2.42, 2.43, 2.44, 2.45, 2.46, 2.47, 2.48, 2.49, 2.50, 2.51, 2.52, 2.53, 2.54, and 2.55.

[2] On the Caverio chart, see note 95 in Chap. 1.

[3] Waldseemüller's use of Caverio's chart is well demonstrated by Joseph Fischer and Franz Ritter von Wieser, *Die älteste Karte mit dem Namen Amerika* (see note 14 in Front Matter), pp. 26–29.

[4] On the encyclopedic character of medieval *mappaemundi*, see Richard Uhden, "Zur Herkunft und Systematik der mittelalterlichen Weltkarten," *Geographische Zeitschrift* 37 (1931), pp. 321–340, esp. 321; Peter Barber, "Visual Encyclopedias," *The Map Collector* 48 (1989), pp. 2–8; and especially Margriet Hoogvliet, "Mappae mundi and Medieval Encyclopaedias: Image versus Text," in Peter Binkley, ed., *Pre-Modern Encyclopaedic Texts: Proceedings of the Second COMERS Congress, Groningen, July 1–4, 1996* (Leiden: Brill, 1997), pp. 63–74.

C. Van Duzer, *Henricus Martellus's World Map at Yale (c. 1491)*, Historical Geography and Geosciences,
https://doi.org/10.1007/978-3-319-76840-3_2

Fig. 2.1 World map by Nicolo de Caverio, c. 1504 (Paris, Bibliothèque nationale de France, Cartes et plans, SH archives 1). (By permission of the Bibliothèque nationale de France)

Opportunities to compare two important maps made around the turn of the sixteenth century, one of which was used as a source for the other, do not arise frequently, and this opportunity will provide insights into how Waldseemüller worked and how he made use of his sources, a type of information that we have about very few other cartographers from this period.

One other point should be reiterated before beginning the exposition of the legends on Martellus's map. I will show that Waldseemüller was using as a source a map very similar indeed to the Yale Martellus map, but there are also significant differences between the texts on the two maps. While some of these differences are certainly the result of Waldseemüller simply choosing not to follow Martellus, the character of some of the differences is such that it seems safe to conclude that Waldseemüller was using a different version of Martellus's large world map.

LEGENDS OUTSIDE OF THE MAP PROPER

Legends Describing the Winds

Martellus's image of the world on his Yale map is surrounded by twelve wind-heads. This decorative element is common on Ptolemaic world maps, going back to the earliest surviving manuscript of Ptolemy's *Geography* that has maps, namely, Vatican City, Biblioteca Apostolica Vaticana, MS Urb. graec. 82, made c. 1300,[5] and they appear on the world maps in both of the manuscripts of the *Geography* that Martellus made (see Fig. 1.1) and also on the Martellus-Rosselli map (Fig. 1.6).[6] However, on the Yale map, Martellus includes with each wind-head a descriptive text, and this element does not appear on Ptolemaic world maps or on the Martellus-Rosselli map. This element is also rare on nautical charts and in fact is much more common on medieval *mappaemundi*. The texts are thus an unusual feature of the map, and although Waldseemüller included wind-heads on his 1507 map, he did not follow Martellus in including texts about the winds (if the version of Martellus's map that Waldseemüller

[5]This manuscript is reproduced in facsimile in Fischer, *Claudii Ptolemaei Geographiae, Codex Vrbinas Graecvs 82* (see note 8 in Chap. 1). Wind-heads also appear on a number of medieval *mappaemundi*, and the earliest appearance of wind-heads on maps that I know of is on a map described in a poem by Theodulf (c. 760–821), specifically near the end of his poem 47, "On Another Picture, in which the Likeness of the Earth was Represented in the Form of a Circle": the Latin text is edited in Ernst Dümmler, ed., *Poetae latini aevi Carolini* (Berlin: Weidmann, 1881), p. 548; the passage is translated in Alexandrenko, Nikolai A., "The Poetry of Theodulf of Orleans: A Translation and Critical Study," Ph.D. Dissertation, Tulane University, 1970, p. 270; and the Latin and English are quoted in Marcia Kupfer, "Medieval World Maps: Embedded Images, Interpretive Frames," *Word and Image* 10 (1994), pp. 262–288, at 266.

[6]On the Martellus-Rosselli map, there are only eight wind-heads rather than twelve; I am inclined to think that Martellus used the smaller number on that map simply for reasons of space. There is an excellent study of the iconography of wind personifications by Thomas Raff, "Die Ikonographie der mittelalterlichen Windpersonifikationen," *Aachener Kunstblätter* 48 (1978–1979), pp. 71–218, but unfortunately he does not address wind-heads in detail. There is some discussion of wind-heads on maps from the Middle Ages through the sixteenth century in Chet Van Duzer, "A Newly Discovered Fourth Exemplar of Francesco Rosselli's Oval Planisphere of c.1508," *Imago Mundi* 60.2 (2008), pp. 195–201.

was consulting included such texts).[7] In what follows I transcribe and translate the legends about the winds, starting in the west and proceeding clockwise around the map; and then I will discuss the texts about winds on a few other maps for purposes of comparison. Martellus took his texts about the winds, whether directly or indirectly, from Chapter 37 of Isidore of Seville's *De natura rerum*, a manual of astronomy and geography written in the seventh century which enjoyed wide popularity during the Middle Ages.[8]

1. The legend about the west wind (Zephyrus) is almost legible using the 2010 ultraviolet image but may be easily read using a multispectral image (tile C01R01 or C01R02). The text comes from Isidore's *De natura rerum* 37.4, though in the second line Martellus has *inferiori* where Isidore has *interiori*[9]:

   favonius sive Zephyrus quartus cardinalis
   ab occidente inferiori flat. Iste hyem[is]
   frigorem gratissima vice relaxat
   floresque producit.

   Favonius or Zephyr is the fourth cardinal wind blowing from low in the west. This wind mitigates the cold of winter with a welcome change and produces flowers.

2. The legend about the wind in the west-southwest, along the left edge of the map. The text is legible in the 2010 ultraviolet image but is much easier to read in the multispectral image (tile C01R04). The source is the same section of Isidore's *De natura rerum*[10]:

   Affricus qui dicitur Lips ex zephyri
   dextro latere intonans: hic generat t[em]
   pestates et pluvias facit & nubium col

[7] Waldseemüller and Matthias Ringmann discuss the winds in Chapter 8 of their *Cosmographiae introductio*, published to accompany the 1507 map, but the sources for the material about the winds there are Roman poets, rather than Isidore's *De natura rerum*, so there is no apparent influence of Martellus's map in that work. For the passage see Martin Waldseemüller, *The Cosmographiae introductio of Martin Waldseemüller in Facsimile: Followed by the Four Voyages of Amerigo Vespucci, with their Translation into English*, trans. Joseph Fischer and Franz von Wieser, ed. Charles George Herbermann (New York: United States Catholic Historical Society, 1907), pp. xxv–xxviii (Latin) and 64–67 (English); and John W. Hessler, *The Naming of America: Martin Waldseemüller's 1507 World Map and the Cosmographiae introductio* (Washington, DC: Library of Congress, 2008), pp. 95–99.

[8] Isidore's *De natura rerum* was published in Augsburg by Günther Zainer in 1472 (under the title *De responsione mundi et de astrorum ordinatione*), so Martellus could have consulted the work in a printed edition rather than in a manuscript. The standard modern edition is Isidore of Seville, *Traité de la nature*, ed. and trans. Jacques Fontaine (Bordeaux: Féret, 1960). There is an English translation as Isidore of Seville, *On the Nature of Things*, trans. Calvin B. Kendall and Faith Wallis (Liverpool: Liverpool University Press, 2016).

[9] Isidore, *De rerum natura*, 37.4, has *Quartus est cardinalis Zephyrus, qui et Favonius ab Occidente interiori flans. Iste hiemis rigorem gratissima vice relaxat, flores producit.* See Fontaine's edition and translation into French in *Traité de la nature* (see note 8 in Chap. 2), pp. 296–297.

[10] Isidore, *De rerum natura*, Chapter 37.4, gives *Africus, qui dicitur Lips ex Zephyri dextro latere intonans: hic generat tempestates, et pluvias, et facit nubium collisiones, et sonitus tonitruorum, et crebrescentium fulgorum visus, et fulminum impulsus.* See Fontaine, *Traité de la nature* (see note 8 in Chap. 2), pp. 296–297.

lisiones et sonitus tonitruorum et crebr
escentium fulgurum visus et fulminum impulsus.
The African wind which is also called Libs, which thunders on the right-hand side of Zephyr. This wind generates storms and causes rain and collisions of clouds, rumbling in the heavens, the sight of very frequent lightning, and bursts of thunder.

3. The legend about the wind in the south-southwest, along the bottom edge of the map, read using an ultraviolet image and multispectral image (tile C02R05); the source is Isidore's *De rerum natura*[11]:

Libonotus est ventus
temperatus calidus
a sinistris austri spirans.
The Libonotus is a temperate warm wind blowing from the left of the South Wind.

4. The legend about the south wind. Some of the paint has flaked off this part of the map, leaving the legend impossible to decipher. Even with multispectral images (tile C05R05), just a few letters of the legend are legible; those letters are consistent with a straightforward disposition of the relevant text from Isidore *De natura rerum* 37.3[12] into five lines, but it is impossible to say more. One interesting detail that the multispectral images reveal is that the painter first drew the lower edge of this banner too high and then had to redraw it lower to make room for the text.
5. The legend about the south-southeast wind (Euronotus), along the bottom edge of the map. The text and even the outlines of the cartouche are invisible in natural light; the cartouche does appear in ultraviolet images, but the words are illegible; however, most of the words are just barely legible in the multispectral images (tile C09R05). Martellus labels the wind *Euronotus*, but in his brief text about the wind, which he again borrows from Isidore's *De natura rerum*,[13] he uses a name that Isidore says is synonymous:

[ ] Eurusauster est calidus
ventus a dextro austri
The South-South-East wind is a warm wind [that blows] from the right of the South Wind.

[11] Isidore, *De rerum natura* 37.3, has *Libonotus, vel Austroafricus ventus est temperatus, calidus a sinistris Austri spirans*. See Fontaine, *Traité de la nature* (see note 8 in Chap. 2), pp. 296–297.

[12] Isidore, *De natura rerum* 37.3 has *Auster, plagae meridianae cardinalis, qui et Notus, ex humili flans, humidus, calidus atque fulmineus, generans largas nubes, et pluvias laetissimas, solvens etiam flores*. See Fontaine, *Traité de la nature* (see note 8 in Chap. 2), pp. 296–297.

[13] Isidore, *De natura rerum* 37.3: *Euronotus. Idem enim Euronotus qui et Euroauster…. Euroauster, calidus ventus, a dextris intonat Austri*. See Fontaine, *Traité de la nature* (see note 8 in Chap. 2), pp. 296–297.

6. The legend about the east-southeast wind (Euroauster). The cartouche is visible in both the natural light and ultraviolet images, but the words are only legible in the multispectral images (tile C11R04):

   Hic eurusauster ex sinistro
   latere subsolani orientem nu
   bibus irrigat calidus [a dextris]
   Intonat austeri

   Here the east-south-east wind [blows] from the left-hand side of the East Wind. It irrigates the East with clouds and is warm, thundering from the right of the South Wind.

   The text comes from Isidore's *De natura rerum*, but in this case, Martellus has combined text from the description of Eurus, the east wind, in 37.2,[14] with Isidore's description of the of the Euroaster in 37.3.[15] It is not clear why he would have done so.

7. The legend about the east wind (Subsolanus), in a cartouche that is oriented vertically. Parts of the first two lines are legible in an ultraviolet image of the map, while the whole legend is legible in a multispectral image (tile C11R02), albeit with considerable difficulty. The source is most likely Isidore's *De natura rerum*[16]:

   Secundus ventorum cardinalis
   Subsolanus est ab ortu intonat
   solis est temperatus atque suavis

   The second cardinal wind is the East Wind, which blows from the sunrise; it is temperate and agreeable.

8. The legend about the east-northeast wind, in the upper right part of the map, is legible in the ultraviolet image and also in the multispectral image (tile C09R01). In this case Martellus edited the text that he found in Isidore (*De natura rerum* 37.2)[17] rather more than in other examples on this map:

   Cecias siue Vulturnus dext
   erior subsolani frigidus valde
   et cuncta disolvit atque desicat.

   The Cecias or Vulturnis [is] to the right of the Subsolanus [and is] very cold and destroys and dries everything.

[14] Isidore, *De natura rerum* 37.2 has *Eurus, ex sinistro latere veniens subsolani, orientem nubibus irrigat.* See Fontaine, *Traité de la nature* (see note 8 in Chap. 2), pp. 294–295.

[15] Isidore, *De natura rerum* 37.3 has *Euroauster, calidus ventus, a dextris intonat Austri. Libonotus, vel Austroafricus ventus est temperatus, calidus a sinistris Austri spirans.* See Fontaine, *Traité de la nature* (see note 8 in Chap. 2), pp. 296–297.

[16] Isidore, *De natura rerum* 37.2 has *Secundus ventorum cardinalis Subsolanus, qui et Apeliotes. Hic ab ortu solis intonat, et est temperatus.* See Fontaine, *Traité de la nature* (see note 8 in Chap. 2), pp. 294–295.

[17] Isidore, *De natura rerum* 37.2 has *Vulturnus ipse, qui et Caecias vocatur, dexterior Subsolani. Hic dissolvit cuncta atque desiccat.* See Fontaine, *Traité de la nature* (see note 8 in Chap. 2), pp. 294–295.

9. The legend about the north-northeast wind is partly legible in an ultraviolet image of the map but may be easily read in the multispectral image (tile C07R01). Martellus's source was Isidore's *De natura rerum*,[18] which again the cartographer has edited rather more than in most of the other cases on this map:

Aquilo qui et boreas gelidus
atque siccus ex alto flans
et sine pluvia qui non
discutit nubes sed stringit

Aquilus which [is] also [called] Boreas [is] icy cold and dry, blowing from on high without rain; it does not disperse the clouds, but presses them together.

10. The legend about the north wind. It is possible to read a few words of the legend in the natural light and ultraviolet images, but the multispectral images reveal the text clearly (tile C06R01). The text comes from Isidore's *De natura rerum*[19]:

Primus ventorum cardinalis
Septentrio frigidus et nivalis
flat rectus ab axe et facit arida
frigora et siccas nubes Et Aparctias dict[us]

The first of the cardinal winds is the North Wind, which is cold and snowy and flows directly from the North and causes an arid cold and dry clouds, and is also called Aparctias.

11. The legend about the north-northwest wind, read using a natural light image and multispectral images (tiles C05R01 and C06R01); the source is Isidore's *De natura rerum*[20]:

Circius qui et thracias hic a dextris
Septentrionis intonans facit nive
m et grandinum coagolationes.

The Circius which is also [called] Thracias thunders here from the right of the North Wind; it makes snow and hailstones.

12. The legend about the west-northwest wind is almost legible using natural light and an ultraviolet image but is more readily legible in the multispectral images (tiles C02R01 and C03R01); the source is Isidore, *De natura rerum*, 37.4–37.5, but where Isidore correctly gives the other name of the wind as *corus*, Martellus mistakenly gives *eurus*[21]:

[18] Isidore, *De natura rerum* 37.1 has *Aquilo ventus qui et Boreas vocatur, ex alto flans, gelidus atque siccus, et sine pluvia, qui non discutit nubes, sed stringit; unde et non immerito diaboli formam induit, qui iniquitatis frigore gentilium corda constringit.* See Fontaine, *Traité de la nature* (see note 8 in Chap. 2), pp. 294–295.

[19] Isidore, *De natura rerum*, 37.1 has *Primus ventorum cardinalis, Septentrio, frigidus et nivalis, flat rectus ab axe, et facit arida frigora et siccas nubes. Hic et Aparctias.* See Fontaine, *Traité de la nature* (see note 8 in Chap. 2), pp. 294–295.

[20] Isidore, *De natura rerum* 37.1 has *Circius que thracias hic a dextris septentrionis intonans facit nives et grandinum coagulationes.* See Fontaine, *Traité de la nature* (see note 8 in Chap. 2), pp. 294–295.

[21] Isidore, *De natura rerum* 37.4-5 has *Corus, qui et Argestes, ex sinistra parte Favonii spirans, eo flante, in Oriente nubila sunt, in India serena. Quosdam autem Tranquillus proprios locorum flatus propriis appellat vocabulis.* See Fontaine, *Traité de la nature* (see note 8 in Chap. 2), pp. 296–297.

*Argestes qui & eurus et [for ex] sinistro [*sic*] parte*
fauoni aspirans, eo flante in orientem
nubila sunt serena quosdam autem
tranquillus proprios locorum flatus certis
appellat vocabulis.

The Argestes which [is] also [called] Eurus blows from the left side of Favonius, and when it blows there are placid clouds in the East. But Suetonius calls some local winds by the names used in those areas.

As mentioned earlier, I know of only one nautical chart that has legends describing the winds, and that is the 1367 nautical chart of the Pizzigani brothers (Parma, Biblioteca Palatina, Carta nautica no. 1612),[22] which is an elaborately decorated luxury chart, and in that case the legends describing the winds do not come from Isidore's *De rerum natura.*[23]

There are closer analogies to Martellus's wind legends among medieval *mappaemundi.* In the early twelfth-century autograph manuscript of Lambert of St. Omer's *Liber floridus* (Ghent, Universiteitsbibliotheek, MS 92), there is a diagram of the winds with a T-O *mappamundi* at the center (f. 24r), and the legends describing the winds come from Isidore's *De rerum natura.*[24] Similar diagrams appear in the Wolfenbüttel, Genoa, Chantilly, and 's-Gravenhage manuscripts of Lambert's work.[25] The legends about the winds on the late thirteenth-century Ebstorf *mappamundi* do not come from Isidore,[26] while the wind legends on the Hereford

[22]The 1367 Pizzigani chart is in Parma, Biblioteca Palatina, Carta nautica no. 1612, and there is a good hand-drawn facsimile of the chart in Edme-François Jomard, *Les monuments de la géographie* (Paris: Duprat, 1842–1862), nos. 44–49; and a photographic reproduction in Cavallo, *Cristoforo Colombo e l'apertura degli spazi* (see note 42 in Chap. 1), vol. 1, pp. 432–433. There is a good digital reproduction of the chart in Pujades i Bataller, *Les cartes portolanes*, on the accompanying CD, number C13. The Library of Congress has a hand-painted copy of the map made by Agostinho Sardi in Parma in 1802, which is very briefly described in Walter W. Ristow and R. A. Skelton, *Nautical Charts on Vellum in the Library of Congress* (Washington, DC: The Library, 1977), p. 2. There is also a hand-painted copy of the map made in 1827, which is in Vienna, Österreichische Nationalbibliothek, Cod. Ser. Nov. 4676.

[23]For transcriptions of the legends about the winds on the Pizzigani chart, see Mario Longhena, "La carta dei Pizigano del 1367 (posseduta dalla Biblioteca Palatina di Parma)," *Archivio storico per le province Parmensi*, series 4, vol. 5 (1953), pp. 25–130, esp. 51.

[24]See Lambert of St. Omer, *Lamberti S. Avdomari Canonici Liber Floridvs. Codex authographus bibliothecae universitatis Gandavensis*, ed. Albertus Derolez (Ghent: In aedibus Story-Scientia, 1968), which includes a color reproduction of the map in question on p. 49, and a transcription of the legends about the winds in the section at the end of the volume, p. 9. The diagram is also reproduced, with brief commentary, in Barbara Obrist, "Wind Diagrams and Medieval Cosmology," *Speculum* 72.1 (1997), pp. 33–84, at p. 55; there is more detailed discussion of the diagram in Karen De Coene, "Van waar de wind waait. Het religieuze wereldbeeld van Lambertus van Sint-Omaars," *Trajecta* 18 (2009), pp. 3–22.

[25]See Destombes, *Mappemondes* (see note 10 in Front Matter), p. 113, where this diagram is indicated as number II, and pp. 115–116 for details of the manuscripts.

[26]The legends about the winds on the Ebstorf *mappamundi*, starting in the west and proceeding clockwise, may be found in Hartmut Kugler, *Die Ebstorfer Weltkarte* (Berlin: Akademie Verlag, 2007), (1) vol. 1, p. 142, no. 57.35, and vol. 2, p. 316; (2) vol. 1, p. 148, no. 60.3, and vol. 2, p. 328; (3) vol. 1, p. 126, no. 49.10, and vol. 2, p. 275; (4) vol. 1, p. 84, no. 28.10, and vol. 2, p. 156; (5) vol. 1, p. 56, no. 14.7, and vol. 2, p. 104; (6) vol. 1, p. 40, no. 6.3, and vol. 2, p. 84; (7) vol. 1, p. 36, no. 4.3, and vol. 2, p. 79; (8) vol. 1, p. 32, no. 2.2, and vol. 2, p. 75; (9) vol. 1, p. 58, no. 15.1, and vol. 2, p. 105; (10) vol. 1, p. 72, no. 22.3, and vol. 2, p. 129; (11) vol. 1, p. 114, no. 43.12, and vol. 2, p. 237; and (12)

mappamundi combine information from Isidore's *De natura rerum* 37 and his *Etymologiae* 13.11.[27] The presence of this type of wind legends in a large and elaborately decorated nautical chart and two of the most ambitious medieval *mappaemundi*, but in few other works, confirms the ambitious nature of Martellus's map at Yale.

Text in the Lower Left Corner

Almagià in his report on the Yale Martellus map writes that "The lower portion of the map contains two rectangular panels, the one on the left has a legend which was glued over the rectangular space (or over a pre-existing legend) which is of no interest."[28] It would certainly be interesting to know what is beneath the glued-on panel and whether, for example, this revision was made merely to correct an error, but the multispectral images of the map made in 2014 unfortunately do not reveal any of the undertext. But the legend that is visible certainly is of interest, as comparison with the legend in the corresponding position on Waldseemüller's 1507 map shows some of Waldseemüller's borrowing from Martellus. Here is the legend on Martellus's map, which is one of the very few texts on the map that is easy to read in natural light, followed by an English translation:

> *ET SI STRABO AC PTHOLEMEUS ET PLERIQUE VETERES DESCRIBENDI ORBIS FUERE STUDIOSISSIMI NOVORUM TAMEN DILIGENTIA QUEDAM AB EIS TANQUAM INCOGNITA PRETERMISSA ADINVENIT QUE NOS HUIC PICTURE AD VERAM LOCORUM SCIENTIAM EXPRIMENDAM STUDIOSE IUNXIMUS.*
>
> *Although Strabo and Ptolemy and many ancient authors were very assiduous in describing the world, nonetheless the diligence of recent men has discovered things omitted and apparently unknown to them, which we carefully include in this image in the interest of expressing the true knowledge of places.*

This text again shows Martellus's interest in new geographical discoveries, and it is worth emphasizing that Martellus's attitude toward classical authorities, and his willingness to declare their information incomplete, is to be contrasted with the very deferential attitude toward classical authors shown by Martellus's former colleague and perhaps master Donnus Nicolaus Germanus.[29] It is interesting that Martellus mentions Strabo and Ptolemy together here, as Strabo had been held to be of higher authority earlier in the Renaissance.[30]

---

vol. 1, p. 140, no. 56.5, and vol. 2, p. 308.

[27] On the wind legends on the Hereford *mappamundi*, see Scott D. Westrem, *The Hereford Map: A Transcription and Translation of the Legends with Commentary* (Turnhout: Brepols, 2001), pp. 12–19, nos. 20–31.

[28] Roberto Almagià, "Worldmap by Henricus Martellus at Berne," June, 1960 (see note 2 in Front Matter), p. 2.

[29] On Nicolaus Germanus's deferential attitude toward classical authorities, see note 43 in Chap. 1.

[30] See Milton Vasil Anastos, "Pletho, Strabo, and Columbus," *Annuaire de l'Institut de Philologie ed d'Histoire Orientales et Slaves* 12 (1952), pp. 1–18 = *Mélanges Henri Grégoire*, vol. 4 (Brussels: Secrétariat des éditions de l'Institut, 1953); reprinted in Milton V. Anastos, *Studies in Byzantine Intellectual History* (London: Variorum Reprints, 1979), article XVII, esp. pp. 13–18.

Here follows the corresponding legend on Waldseemüller's 1507 map; Waldseemüller's is longer, in part because he discusses the discovery of the New World in the beginning of the text,[31] which Martellus could not do, as the New World had probably not been discovered when he made his map. But Waldseemüller had certainly studied Martellus's map, for he copies the final phrase of Martellus's legend verbatim (underlined for emphasis):

> *Terrarum insularumque variarum generalis descriptio. etiam quarum vetusti non meminerunt auctores nuper ab anno domini .1497. usque ad .1504. bis geminis navigationibus in mare discursis inventarum. duabus per fernandum castille reliquis vero duabus in australi ponto per dominum manuelem portugallie serenissi[mi] reges americo vesputio uno ex naucleris nauiumque prefectis precipue multorum etiam locorum quorum nulla erat noticia. que nos huic picture ad veram locorum scientiam exprimendam studiose junximus.*

And an English translation[32]:

> *A general delineation of the various lands and islands, including some of which the ancients make no mention, discovered lately between 1497 and 1504 in four voyages over the seas, two commanded by Fernando of Castile, and the other two, to the southern sea, by Manuel of Portugal, most serene monarchs, with Amerigo Vespucci as one of the navigators and officers of the fleet; and especially a delineation of many places hitherto unknown, which we carefully include in this image in the interest of revealing the true knowledge of places.*

Thus even when Waldseemüller was writing about a subject that Martellus's had not addressed in the text in the corresponding position on his map, Waldseemüller consulted that corresponding text and borrowed from it.

Text in the Lower Right Corner

Small parts of this text are legible in natural light, ultraviolet, and infrared images of the map, but the whole thing may be easily read in the multispectral images (tiles C10R04, C11R04, C10R05, and C11R05) (see Fig. 2.2). The legend is of interest for several reasons. First, in the opening Martellus emphasizes that the world is surrounded by the ocean, and we have seen several pieces of evidence that indicate that this subject was of interest to him: in the manuscripts of his *Insularium* and in the Magliabechiano

[31] In fact the first part of Waldseemüller's legend comes from the preface to the edition of Vespucci's voyages in the *Cosmographiae introductio*: see Waldseemüller, *The Cosmographiae introductio of Martin Waldseemüller in Facsimile* (see note 7 in Chap. 2), pp. xlv (Latin) and 88 (English). And incidentally the phrase "quarum vetusti non meminerunt autores" from Waldseemüller's text reappears in the title of Waldseemüller's *Carta marina* of 1516, which runs *Carta marina navigatoria, Portugallenses navigationes atque tocius cogniti orbis terre marisque formam naturamque, situs et terminos nostris temporibus recognitos et ab antiquorum traditione differentes, eciam quorum vetusti non meminerunt autores, hec generaliter indicat*, that is, "This marine nautical chart indicates generally the navigations of the Portuguese, and the form and nature of the whole known sphere of land and of sea, the places and boundaries discovered in our times and differing from the tradition of the ancients, as well as those which ancient authors did not mention."

[32] The English translation is from Hessler, *The Naming of America* (see note 7 in Chap. 2), p. 17, with some modifications. The legend is also transcribed, translated into English, and discussed in Gaetano Ferro, Luisa Faldini, Marica Milanesi, et al., *Columbian Iconography*, ed. Gianni Eugenio Viola (Rome: Istituto polografico e Zecca dello Stato, Libreria dello Stato, 1992), pp. 462–463.

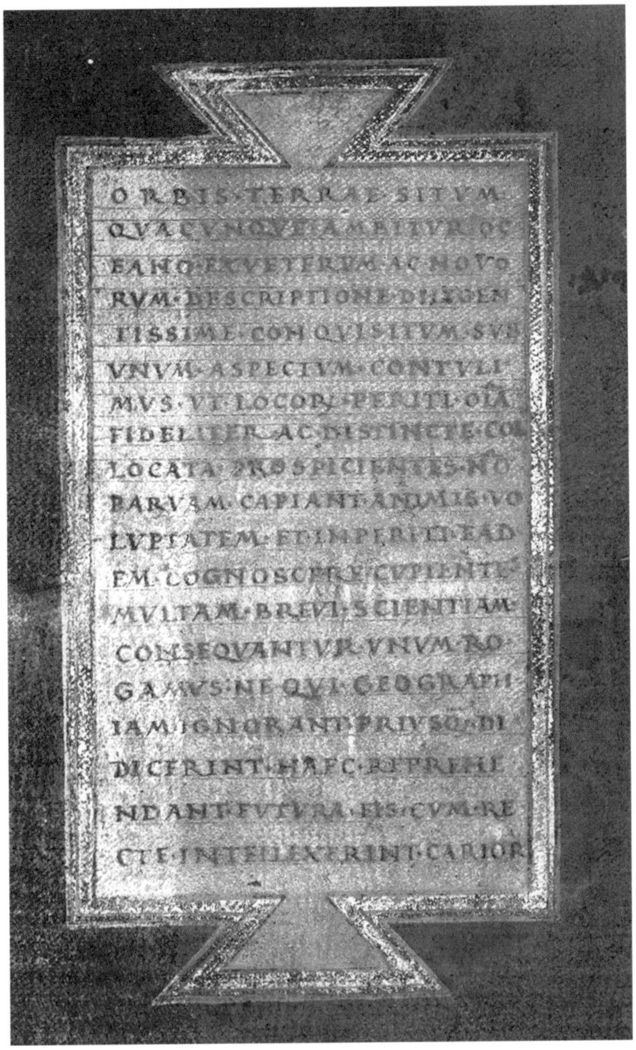

Fig. 2.2 Multispectral image of the text box in the lower right-hand corner of the Yale Martellus map. (Image by Lazarus Project/MegaVision/RIT/EMEL, courtesy of the Beinecke Rare Book and Manuscript Library)

manuscript of Ptolemy, Martellus cites Piccolomini's discussion in his *Historia rerum ubique gestarum* of whether the inhabited part of the world is surrounded by the ocean, and the title of the Martellus-Rosselli map begins with the same idea: *forma uniuersalis totius quod oceano mari ambitur*.... Also, though Martellus was far from the first to talk about the geographical knowledge of the ancients being supplemented or corrected by that of modern explorers, given Waldseemüller's use of Martellus's map, it is tempting to think that Martellus's words on this subject in both the text about to be cited and the one in the lower left-hand corner of the map just examined inspired Waldseemüller to include his portraits of Ptolemy and Vespucci, representatives of ancient and modern geographical knowledge, at the top of his 1507 map. And finally, as I will show below, Waldseemüller borrowed words and phrases from the following legend in the texts in both the upper right and lower right corner of his map. Here is Martellus's text, followed by an English translation:

ORBIS TERRAE SITUM
QUACUNQUE AMBITUR OC
EANO EX VETERUM AC NOVO
RUM DESCRIPTIONE[M] DILIGEN
*TISSIME CONQUISITUM SUS [*sic, *for* SUB*]*
UNUM ASPECTUM CONTULI
MUS UT LOCORUM PERITI OMNIA
FIDELITER AC DISTINCTE COL
LOCATA PROSPICIENTES NON
PARVAM CAPTANDAM ANIMIS VO
LUPTATEM ET INPERITI EAD
EM COGNOSCERE CUPIENTES
MULTAM BREVI SCIENTIAM
CONSEQUANTUR UNUM RO
GAMUS NE QUI GEOGRAPH
IAM IGNORANT PRIUSQUAM DI
DICERINT HAEC REPREHE
NDANT FUTURA EIS CUM RE
CTE INTELLEXERINT CARIOR[A]

The position of the inhabited world, which is surrounded on all sides by the ocean, we have united in one view from very diligently selected maps of the ancients and moderns, so that those who are knowledgeable about places, seeing everything faithfully and carefully located, will have no small pleasure in their souls, while those who are inexperienced but wish to learn about places can acquire much knowledge in a short time. We ask one thing: that those who are ignorant of geography will learn these things before criticizing—things that will be more valuable to them when they understand them correctly.

Here follow first the legend from Martellus again, with some words and phrases that are echoed by Waldseemüller underlined; second, the legend from the upper right corner of Waldseemüller's map, with words and phrases that echo Martellus's legend underlined; and third, an English translation of Waldseemüller's legend. Here is Martellus's legend:

ORBIS TERRAE SITUM
QUACUNQUE AMBITUR OC
EANO EX VETERUM AC NOVO
RUM DESCRIPTIONE[M] DILIGEN
*TISSIME CONQUISITUM SUS [*sic, *for* SUB*]*
UNUM ASPECTUM CONTULI
MUS UT LOCORUM PERITI OMNIA
FIDELITER AC DISTINCTE COL
LOCATA PROSPICIENTES NON
PARVAM CAPTANDAM ANIMIS VO
LUPTATEM ET INPERITI EAD
EM COGNOSCERE CUPIENTES
MULTAM BREVI SCIENTIAM
CONSEQUANTUR UNUM RO
GAMUS NE QUI GEOGRAPH
IAM IGNORANT PRIUSQUAM DI

DICERINT HAEC REPREHE
NDANT FUTURA EIS CUM RE
CTE INTELLEXERINT CARIOR[A]

Now the legend in the upper right corner of Waldseemüller's map:

TIPUM ORBIS GENERALEM DESCRIBENDO: VE-
TERUM INVENTA PONERE/ ET EA QUE A NEO-
TERICIS INTERIM REPERTA SUNT (SICUT EST
CATAIA REGIO) CONIUNGERE PLACUIT. UT
TALIUM RERUM STUDIOSI DUM VARIA COGNO-
SCERE CUPIUNT VOTORUM COMPOTES LABO-
RI NOSTRO SINT GRATI PLAERAQUE OMNIA
TAM PASSIM COGNITA QUAM NOVITER LUS-
TRATA DILIGENTER AC DISTINCTE SUB UNO
ASPECTU COLLOCATA PROSPICIENTES.

And now an English translation of Waldseemüller's legend[33]:

In describing the general appearance of the whole world, it has seemed best to put down the discoveries of the ancients, and to add what has since been discovered by the moderns, for instance, the land of Cathay, so that those who are interested in such matters and wish to find out various things may gain their wishes and be grateful to us for our labor, when they see nearly everything that has been discovered here and there, or recently explored, carefully and clearly brought together, so as to be seen at a glance.

We see clearly that Waldseemüller had Martellus's text in front of him and used it for inspiration and borrowed words and phrases from it but created something that was his own.

The situation is similar with the legend in the lower right corner of Waldseemüller's map: Waldseemüller, toward the end of his legend, borrows phrases from the end of Martellus's legend in which he asks those who view the map who are not knowledgeable about cartography to refrain from criticizing it.[34] Here follow first Martellus's legend again, with the words that Waldseemüller echoes in the lower right corner of his map underlined; second, the legend from the lower right corner of Waldseemüller's map, with words and phrases that echo Martellus's legend underlined; and third, an English translation of Waldseemüller's legend. Martellus's legend:

ORBIS TERRAE SITUM
QUACUNQUE AMBITUR OC
EANO EX VETERUM AC NOVO
RUM DESCRIPTIONE[M] DILIGEN

[33]The translation of Waldseemüller's legend is from Hessler, *The Naming of America* (see note 7 in Chap. 2), p. 30.

[34]Many medieval and Renaissance authors asked for their readers' indulgence, but it is tempting to think that Martellus was inspired here by a similar request that Donnus Nicolaus Germanus, his former colleague and/or teacher, made in the dedication of some of his manuscripts of Ptolemy's *Geography*, where he writes: "If those who are not altogether ignorant of geography or cosmography, and are in the habit of reading Ptolemy, will compare, with a calm mind, our picture with his, they will certainly think our picture worthy of some praise, instead of blaming it…." For the bibliographic details see note 43 in Chap. 1.

TISSIME CONQUISITUM SUS [sic, *for* SUB]
UNUM ASPECTUM CONTULI
MUS UT LOCORUM PERITI OMNIA
FIDELITER AC DISTINCTE COL
LOCATA PROSPICIENTES NON
PARVAM CAPTANDAM ANIMIS VO
LUPTATEM ET INPERITI EAD
EM COGNOSCERE CUPIENTES
MULTAM BREVI SCIENTIAM
CONSEQUANTUR UNUM RO
GAMUS NE QUI GEOGRAPH
IAM IGNORANT PRIUSQUAM DI
DICERINT HAEC REPREHE
NDANT FUTURA EIS CUM RE
CTE INTELLEXERINT CARIOR[A]

The legend from the lower right corner of Waldseemüller's map:

LICET PLAERIQUE VETERUM DESCRIBENDI TERRA-
RUM ORBIS STUDIOSISSIMI FUERINT: NON TAMEN
PARUM IPSIS EISDEM INCOGNITA MANSERUNT, SI-
CUT EST IN OCCASU AMERICAE: AB EIUS NOMINIS
INVENTORE DICTA, QUE ORBIS QUARTA PARS PU-
TANDA EST. SICUT ET VERSUS MERIDIEM APHRICE
PARS: QUAE SEPTEM PENE GRADIBUS CITRA CAPRI-
CORNUM INCIPIENS/ULTRA TORRIDAM ZONAM ET
EGOCERI TROPICUM AD AUSTRUM LATISSIME PRO-
TENDTUR. SICUT QUOQUE IN TRACTU ORIENTA-
LI REGIO CATAIAE/ET QUICQUID INDIAE MERIDIO-
NALIS ULTRA CENTESIMUM ET OCTOGESIMUM LON-
GITUDINIS GRADUM EST SITUM. QUAE NOS PRIO-
RIBUS OMNIA ADJUNXIMUS: UT ISTIUSCEMODI RE-
RUM AMATORES/QUAECUNQUE SUB HANC DIEM
NOBIS PATENT OCULIS INTUENTES/DILIGENTIAM
NOSTRAM POBENT. ID AUTEM UNUM ROGAMUS
UT RUDES ET COSMOGRAPHIAE IGNARI HAEC NON
STATIM DAMNENT ANTEQUAM DIDICERINT CHA-
RIORA IPSIS HAUD DUBIE POST CUM INTELLEXE-
RINT FUTURA.

And my translation[35]:

Many of the ancients were very zealous in describing the world; however, no small part of the world remained unknown to them, as is the case with America in the West, whose name derives from that of her discoverer, and which should be considered a fourth part of the world. This is also the case in the South, with the part of Africa that begins almost seven degrees north of the Tropic of Capricorn and extends far to the south across the Torrid Zone and the Tropic of

[35]This legend is illustrated, transcribed, and translated into English in Gaetano Ferro, Luisa Faldini, Marica Milanesi, et al., *Columbian Iconography*, ed. Gianni Eugenio Viola (Rome: Istituto polografico e Zecca dello Stato, Libreria dello Stato, 1992), pp. 456–457, and is also translated into English by Hessler, *The Naming of America* (see note 7 in Chap. 2), p. 34.

Capricorn. It is also the case in the East, with the land of Cathay and whatever part of southern India lies beyond the meridian at 180 degrees. All of these we have added here to the findings of the ancients, so that lovers of things of this nature, seeing what has been revealed to our eyes, should approve of our diligence. This alone we ask, that the uneducated and those ignorant of cosmography not condemn these things before they learn what will certainly be more valuable to them when they understand it.

LEGENDS IN THE OCEANS

Legends in the Northern Ocean

On Martellus's map the text in the cartouche immediately below the windhead of the north wind describes that wind; that legend was transcribed and discussed above. To the left of that cartouche, there is another cartouche, below that cartouche an island, and below the island a smaller cartouche. The island and the two cartouches are closely connected, and I will discuss them together. The text on all three is invisible in natural light images of the map, but some of it is legible in the multispectral images (tile C05R02). The legible parts of the large upper cartouche run:

Iste oce [ ]
quod in lingua scithica [ ]
pretatur [ ]
eas montibus est Insula B[annona]
qua colligitur eletrum [sic]*[ ]*
[ ]ambrum[ ]
This ocean...which in the Scythian language.... [is inter]preted... mountains there is an island.... where amber is gathered.... amber....

The text on the island reads:

bannona insula septentrionalis
Bannona, a northern island

In the small cartouche south of the island, just the first line and part of the second line are legible. The legible words read:

Hec insula distat ab continenti
[ ] navigatione
This island is distant from the continent.... by boat

The source of this island and the related descriptive texts is a passage in Pliny's *Naturalis historia* (4.13.94) in which he describes the Northern Ocean[36]; I underline the parts that relate to the texts just cited on Martellus's map:

[36]The Latin is from Pliny the Elder, *C. Plini Secundi Naturalis historiae libri XXXVII*, ed. Karl Mayhoff (Leipzig: Teubner, 1906–1909), vol. 1, p. 344; and the English translation is from Pliny the Elder, *The Natural History*, trans. John Bostock and H. T. Riley (London: H. G. Bohn, 1855–1857), vol. 1, pp. 339–342 (Book 4, Chapter 27, in the numbering of this edition), but I have replaced Bostock's reading of "Raunonia" for the name of the island with "Baunonia," the reading in the Latin text; in a note, Bostock remarks that "Raunonia" is an emendation of the older reading "Bannomanna."

Exeundum deinde est, ut extera Europae dicantur, transgressisque Ripaeos montes litus oceani septentrionalis in laeva, donec perveniatur Gadis, legendum. insulae complures sine nominibus eo situ traduntur, ex quibus ante Scythiam quae appellatur Baunonia unam abesse diei cursu, in quam veris tempore fluctibus electrum eiciatur, Timaeus prodidit. reliqua litora incerta. signata fama septentrionalis oceani. Amalchium eum Hecataeus appellat a Parapaniso amne, qua Scythiam adluit, quod nomen eius gentis lingua significat congelatum.

We must now leave the Euxine to describe the outer portions of Europe. After passing the Riphaean mountains we have now to follow the shores of the Northern Ocean on the left, until we arrive at Gades. In this direction a great number of islands are said to exist that have no name; among which there is one which lies opposite to Scythia, mentioned under the name of Baunonia, and said to be at a distance of the day's sail from the mainland; and upon which, according to Timaeus, amber is thrown up by the waves in the spring season. As to the remaining parts of these shores, they are only known from reports of doubtful authority. With reference to the Septentrional or Northern Ocean; Hecataeus calls it, after we have passed the mouth of the river Parapanisus, where it washes the Scythian shores, the Amalchian sea, the word 'Amalchian' signifying in the language of these races, frozen.

Several authors have discussed Pliny's passage about the island Bauonia,[37] but without shedding any significant light on it. Martellus's is the only cartographic interpretation of this passage in Pliny that I know of; he does not include the island or the associated texts on any of his other world maps, except on the wall map represented in the Bíblia dos Jerónimos (discussed above), where there is an island in this same area.

Significantly, neither this island nor the associated texts appear on Waldseemüller's 1507 map, and this difference between the maps is good evidence that Waldseemüller was consulting a different version of Martellus's map. For Waldseemüller does have a long legend in this same area, and the subject of Waldseemüller's legend, the question of whether the *orbis terrarum* or inhabited part of the world is circumnavigable, is one that interested Martellus greatly, one that he had addressed in texts that accompany the world maps in manuscripts of his *Insularium*. Here is Waldseemüller's legend:

Dubium fuit aliquamdiu vltra Caspium pelagus quidnam esset Enimve[r]o cum Pisici atque ipse Homerus orbem fretu circumfusum esse dixerint/ memorat Cornelius Nepos Q. Metellum Celerem ita retulisse. Cum Gallis proconsul esset Indos quosdam a rege Suaeuorum dono sibi datos unde in eas terras deuenissent requirendo cognosse intempestatum ex Indicis aequoribus abreptos emersosque &c. Tandem in Termaniae littora exijsse. Quo fit vt restet mare.

It was doubtful for some time what sea was beyond the Caspian. Indeed while scientists and Homer himself said that the circle of lands was surrounded by the ocean, Cornelius Nepos reports that Q. Metellus Celer said that when he was proconsul over the Gauls, the king of the Swabians gave him some men from India as a gift, and when he asked whence they had come to those lands, he learned that they had been snatched by a storm from Indian waters and that they had finally arrived on the shores of Germany. The result is that it is still the sea [i.e. the circumfluent ocean is continuous].

[37] Karl Müllenhoff, *Deutsche Altertumskunde* (Berlin: Weidmann, 1890–1908), vol. 1, pp. 476, 481, and 483; Detlef Detlefsen, "Zur Kenntnis der Alten von der Nordsee," *Hermes* 32 (1897), pp. 191–201, at 195; Franz Matthias, *Über Pytheas von Massilia und die ältesten nachrichten von den Germanen* (Berlin: Druck von W. Pormetter, 1901–1902), vol. 1, pp. 30–31; and Fridtjof Nansen, *In Northern Mists: Arctic Exploration in Early Times*, trans. Arthur G. Chater (New York: Frederick A. Stokes Co., 1911), vol. 1, pp. 70 and 98–99

The ultimate source of this information is Pomponius Mela 3.5,[38] and there is a shorter account of the same episode, but with more emphasis on its geographical implications, in Pliny *Naturalis historia* 2.67.170.[39] This same episode is mentioned by Piccolomini in his discussion of whether the ocean can be circumnavigated in his *Historia rerum ubique gestarum*,[40] which is precisely the passage that Martellus cites accompanying the world maps in manuscripts of his *Insularium* and in the Magliabechiano manuscript of Ptolemy's *Geography*.[41] That is, Waldseemüller cites material that

[38]See Pomponius Mela, *Pomponius Mela's Description of the World*, trans. Frank E. Romer (Ann Arbor: University of Michigan Press, 1998), pp. 113–114: "For quite some time it was unclear what lay beyond Caspian Bay, whether it was the same Ocean or a hostile, cold land that extended without a border and without end. But in addition to the natural philosophers and Homer, who all said that the entire known world was surrounded by sea, there is Cornelius Nepos, who is more dependable as an authority because he is more modern. Nepos, however, adduces Quintus Metellus Celer as witness of the fact, and he records that Metellus reported it as follows. When Celer was proconsul of Gaul, certain Indians were presented to him as a gift by the king of the Boii. By asking what route they had followed to reach there, Celer learned that they had been snatched by storm from Indian waters, that they had traversed the intervening region, and they finally they had arrived on the shores of Germany. Ergo, the sea is continuous, but the rest of that same coast is frozen by the unremitting cold and is therefore deserted."

[39]For the discussion of the arrival of Indians on the coast of Germany, see Humphrey Gilbert, *A Discourse written by Sir Humphrey Gilbert Knight, to Proue a Passage by the Northwest to Cathaia, and the East Indies*, in Richard Hakluyt, *The Principal Navigations, Voyages, Traffiques and Discoveries of the English Nation* (Glasgow: J. MacLehose and Sons, 1903–1905), vol. 7, pp. 158–203, esp. Chap. 4, pp. 172–173; Richard Hennig, *Terrae Incognitae: Eine Zusammenstellung und kritische Bewertung der wichtigsten vorcolumbischen Entdeckungsreisen an Hand der darüber vorliegenden Originalberichte* (Leiden: E. J. Brill, 1944–1956), vol. 1, pp. 289–292; J. André, "Des Indiens en Germanie?" *Journal des Savants* 1982, pp. 45–55; Benjamin Olshin, "A Sea Discovered: Pre-Columbian Conceptions and Depictions of the Atlantic Ocean," Ph.D. Dissertation, University of Toronto, 1994, pp. 106–109; Klaus Tausend, "Inder in Germanien," *Orbis Terrarum* 5 (1999), pp. 115–125; and Alexander V. Podossinov, "The Indians in Northern Europe? On the Ancient Roman Notion of the Configuration of Eurasia," in Alexander V. Podossinov, ed., *Periphery of the Classical World in Ancient Geography and Cartography* (Leuven; Prais; Walpole, MA: Peeters, 2014), pp. 133–145. Also, the passage in Pomponius Mela is alluded to in a legend in the Atlantic on the Genoese world map of 1457 (Florence, Biblioteca Nazionale Centrale, Portolano 1); for a transcription and translation of the legend, see Edward Luther Stevenson, *Genoese World Map, 1457* (New York: American Geographical Society and Hispanic Society of America, 1912), pp. 55–56. And Christopher Columbus was very interested in evidence of lands beyond the ocean washed up on sea shores: see Fernando Colón, *The History of the Life and Deeds of the Admiral Don Christopher Columbus, Attributed to his Son Fernando Colón*, ed. Ilaria Caraci Luzzana, trans. Geoffrey Symcox and Blair Sullivan (Turnhout: Brepols, 2004), Chapter 9, pp. 45–49 (English) and 248–251 (Italian).

[40]Piccolomini, *Historia rerum ubique gestarum*, Chapter 2: *Plinius Nepotis testimonio utitur qui Metello Celeri Galliae proconsuli donatos a rege Suevorum Indos astruit qui ex India commercii causa navigantes tempestatibus essent in Germaniam arrepti. Nos apud Ottonem legimus sub imperatoribus Teutonicis Indicum navem et negociatores Indos in Germanico litore fuisse deprehensos quos ventis agitatos ingratis ab orientali plaga venisse constabat: quod accidere minime potuisset si, ut plerique visum est, Septentrionale pelagus innavigabile concretumque esset. A columnis Herculeis Mauritanie atque Hispaniae et Galliarum circuitus....* This chapter is part of Piccolomini's description of Asia, which has been edited and translated into Spanish by Domingo F. Sanz as *Descripción de Asia, Eneas Silvio Piccolomini (Papa Pío II)* (Madrid: Consejo Superior de Investigaciones Científicas, 2010), in which the passage in question is on pp. 106 (Latin) and 107 (Spanish).

[41]Incidentally this passage in Piccolomini's work attracted a marginal comment in Columbus's copy of the book: see the facsimile edition of that copy, published as *Historia rerum: cuyo original se encuentra en la Biblioteca Colombina de Sevilla* (Madrid: Testimonio Compañía Editorial, 1991), f. 2v, and translated as *Historia rerum ubique gestarum del papa Pio II*, trans. Antonio Ramírez de Verger (Madrid: Testimonio Compañía Editorial, 1991), p. 13.

Martellus frequently presents with his world maps, even in a manuscript of Ptolemy's *Geography*, where material from Piccolomini does not belong, in a sense. The combination of the association of this material with Martellus's world maps and Waldseemüller's heavy use of Martellus as a source for the descriptive texts on his map makes it seem likely that this legend on Waldseemüller's world map was inspired by Martellus—specifically, that Waldseemüller was using a version of Martellus's large world map that included a cartouche containing this material.

Legend in the Atlantic Ocean

There is a banner just off the western coast of North Africa and just north of the equator on Martellus's map which unfortunately is illegible in natural, ultraviolet, infrared, and multispectral images of the map (tile C02R02). There is a legend in essentially the same spot on Waldseemüller's map, just a bit further off the coast, and it is tempting to think that Waldseemüller borrowed from Martellus here, particularly as there are no other texts off the west coast of Africa on either map (aside from toponyms), which makes the coincidence of position striking. Waldseemüller's legend, which refers to the Cape Verde islands, reads *Insule Portugalensium inuente tempore herici* [*sic*] *infantis Anno 1472*, "Islands of the Portuguese discovered in the time of the Infante Henry, in the year 1472." There is a problem with the date here: not only were the islands actually discovered c. 1456,[42] but in addition, Henry the Navigator died November 13, 1460, so the islands cannot have been discovered in 1472 and during Henry's time.[43]

[42] And the Cape Verde islands are well delineated in Grazioso Benincasa's atlas of 1468 (i.e., on a map made before 1472) which is in London, British Library Add. MS 6390, f. 9r. The map is reproduced in Henry Yule Oldham, "The Discovery of the Cape Verde Islands," in *Festschrift Ferdinand Freiherrn von Richthofen zum sechzigsten Geburtstag am 5. mai 1893 dargebracht von seinen Schülern* (Berlin: D. Reimer, 1893), pp. 181–195 between pp. 194 and 195; for discussion of this atlas, see Marina Emiliani, "Le carte nautiche dei Benincasa, cartografi anconetani," *Bollettino della Real Società Geografica Italiana* 73 (1936), pp. 485–510, no. 10.

[43] For discussion of the discovery of the archipelago, see Oldham, "The Discovery of the Cape Verde Islands" (see note 42 in Chap. 2); António Brásio, "Descobrimiento, povoamento, evangelização do Arquipélago de Cabo-Verde," *Studia: Revista do Centro de Estudos Historicos Ultramarinos* 10 (1962), pp. 49–97, esp. 52–76; and Armando Cortesão, "Descobrimento e representação das ilhas de Cabo Verde na cartografia antiga," *Memórias da Academia de Ciências de Lisboa, Classe de Ciências* 21 (1976–1977), pp. 229–250; also see A. Teixeira da Mota, "Cinco séculos de cartografia das ilhas de Cabo Verde," *Garcia de Orta, Revista da Junta de Investigaçoes do Ultramar* 9.1 (1961), pp. 11–16 with 24 plates. A manuscript by Valentim Fernandes of 1506–1508 gives the earliest known description of the islands, and his maps are the earliest large-scale maps of the islands: see A. Fontoura da Costa, *Cartas das Ilhas do Cabo Verde de Valentim Fernandes, 1506–1508* (Lisbon: Divisão de Publicações e Biblioteca, Agência Geral das Colónias, 1939), who has good lists of the toponyms of the islands in early documents and maps pp. 61–63 and illustrates the islands in several maps on a chart between pp. 56 and 57; and Valentim Fernandes, *Description de la côte occidentale d'Afrique (Sénégal du Cap de Monte, Archipels)*, eds. Théodore Monod, A. Teixeira da Mota, and Raymond Mauny (Bissau: Centro de Estudos da Guiné Portuguesa 1951), esp. pp. 108–147.

The legend south of the southern tip of Africa, where the Atlantic and Indian Oceans meet, addresses the Portuguese discoveries and was discussed above.

Legends in the Western Indian Ocean

As mentioned earlier, the correspondence in the positions of the five cartouches along the eastern coast of Africa on Martellus's and Waldseemüller's maps is striking and suggests that the content of the legends, in addition to their positions, might be the same.

The northernmost of these five legends on Waldseemüller's map describes the Islands of Men and Women, which are mentioned by Marco Polo[44]; it is located just below two islands labeled *viri* and *femine*. Waldseemüller's highly abbreviated legend reads:

> Anno. d[omini] .1285. i[n] vna ex his insul[is] soli h[ab]itaba[n]t Viri i[n] alia fe[m]i[n]e et sim[ul] co[n]ueniunt semel i[n] a[n]no et su[n]t chr[isti]ani h[abe]nt ep[iscopu]m subiectu[m] scoyre.
>
> *In the year of our Lord 1285, in one of these islands only men lived, and in the other women, and they came together once per year; they are Christians and have a bishop who is subject to Socotra.*

On Martellus's map the corresponding legend is also located just below two islands; even in the multispectral images (tile C05R03), it is not possible to read all of the words, but enough of the text is legible to make it clear that Waldseemüller's text follows that of Martellus very closely:

> *[ ] ex his*
> *Insulis soli habitabant viri*
> *in alia femine et simul conueni*
> *[unt] semel in anno sunt cristiani*
> *habent episcopum subjectum archiepiscopo scoyr[e]*
>
> *...of these islands only men lived, and in the other women, and they came together once per year. They are Christians and have a bishop who is subject to the archbishop of Socotra.*

On both maps the island south of the legend just mentioned is Socotra; the legend on Martellus's map (tile C05R03) reads *Scoyra insula Christianum habet archiepiscopum*, "Socotra, an island of Christians, has

[44] In Prásek's edition of the Naples manuscript of Polo—see Marco Polo, *Marka Pavlova z Benátek, Milion: Dle jediného rukopisu spolu s prilusnym zakladem latinskym*, ed. Justin Václav Prásek (Prague: Nákl. Ceské akademie císare Frantiska Iozefa, 1902)—the chapter on the Islands of Men and Women is Book 3, Chapter 37, pp. 182–183. For discussion of these islands, see *The Book of Ser Marco Polo*, ed. and trans. Henry Yule (New York: C. Scribner's Sons, 1903), vol. 2, pp. 405–406; also see Domenico Silvestri's *De insulis et earum proprietatibus*, an encyclopedia of the world's islands written between 1385 and 1410, which was edited by C. Pecoraro in *Atti della Accademia di scienze, lettere e arti di Palermo* 14.2 (1954), pp. 1–319; and now there is an edition with a Spanish translation by José Manuel Montesdeoca, *Los islarios de la época del humanismo: el 'De Insulis' de Domenico Silvestri, edición y traducción* (La Laguna: Servicio de Publicaciones Universidad de La Laguna, 2004), s.v. "Feminina insula" and "Masculina insula," pp. 117 and 156–157 in Montesdeoca's edition. Paul Pelliot, *Notes on Marco Polo* (Paris: Impr. nationale, 1959–1973), vol. 2, pp. 671–725, has a long section on eastern analogues of the tale.

an archbishop," and Waldseemüller has *Scoyra insula Christiana*, "Socotra, a Christian island."

The text in the next cartouche to the south on Martellus's map is illegible in natural light and ultraviolet images, but it is revealed in multispectral images of the map (tiles C04R03 and C05R03; also on C04R04 and C05R04). It reads:

hic cernitur leviathan draco
marinus qui frequenter
cum ceto pise [sic] *pugnat*
Here is seen the leviathan or sea dragon which frequently fights against the whale fish.

Waldseemüller's legend is very similar indeed: *hic cernitur leuiaton draco marinus qui frequenter contra cetum pugnat*, that is, "Here is seen the leviathan or sea-dragon which frequently fights against the whale." The idea that the leviathan fights against the whale is unusual, and does not appear in medieval bestiaries or in medieval encyclopedias such as Isidore of Seville's *Etymologiae* or Thomas of Cantimpré's *De naturis rerum*, which latter has a substantial section on sea monsters. The story comes from the medieval Muslim author Jorath, whose book about animals is now lost,[45] but his account of the leviathan fighting the whale is preserved in Vincent of Beauvais' *Speculum naturale* (thirteenth century)[46]; Arnold the Saxon's *De floribus rerum naturalium* (thirteenth century)[47]; Guido da Pisa's *Expositiones et glose super Comediam Dantis* (fourteenth century)[48]; the *Hortus sanitatis*, an anonymous illustrated encyclopedia first published in 1491[49]; and in the narrative of the travels of Arnold von Harff to the Holy Land, Egypt, and the Indian Ocean in 1496–1499.[50] Martellus's use of the *Hortus Sanitatis* for legends about other animals on his map makes it very

[45]On Jorath see George Sarton, "Jorach (or Yôrâh), Unknown Author of a Book on Animals," *Isis* 15.1 (Feb., 1931), pp. 171–172; and Isabelle Draelants, "Le dossier des livres sur les animaux et les plantes de Iorach: tradition occidentale et orientale," in Isabelle Draelants, Anne Tihon, and Baudouin van den Abeele, eds., *Occident et Proche-Orient: contacts scientifiques au temps des croisades. Actes du Colloque de Louvain-la-Neuve, 24-25 mars 1997* (Turnhout: Brepols, 2000), pp. 191–276.

[46]Vincent de Beauvais, *Speculum naturale* 20.38, "De Leuiathan," in Vincent of Beauvais, *Speculum quadruplex; sive, Speculum maius* (Graz: Akademische Druck- u. Verlaganstalt, 1964–1965), vol. 1, col. 1480.

[47]Arnold the Saxon, *De floribus rerum naturalium*, Book 2 (*De naturis animalium*), Chapter 8 (*De natura operationis piscium*); see Emil Stange, *Die Encyclopädie des Arnoldus Saxo, zum ersten Mal nach einem Erfurter Codex* (Erfurt: F. Bartholomäus, 1905–1907), vol. 1, p. 64–65.

[48]Guido da Pisa, *Expositiones et glose super Comediam Dantis; or, Commentary on Dante's Inferno*, ed. Vincenzo Cioffari (Albany, NY: State University of New York Press, 1974), p. 659 (on *Inferno* 31.49–57).

[49]The passage on the leviathan fighting the whale is in *Hortus Sanitatis* (Strassburg: Johann Prüss, not after 21 Oct. 1497); the entry for "Leuiathan," which is "De piscibus," Chapter 50, contains just a few details and refers the reader to "De animalibus," Chapter 84, also on "Leuiathan," where the story is related. For bibliography on the *Hortus Sanitatis*, see note 100 in Chap. 1.

[50]Arnold Ritter von Harff, *The Pilgrimage of Arnold von Harff, Knight, from Cologne, through Italy, Syria, Egypt, Arabia, Ethiopia, Nubia, Palestine, Turkey, France and Spain, Which He Accomplished in the Years 1496 to 1499*, trans. Malcolm Letts (London: Hakluyt Society, 1946) (= Works Issued by the Hakluyt Society, 2nd ser., no. 94), p. 158

likely that he used the same book as his source for his legend here about the fight between the leviathan and whale.[51]

The next legend to the south on Martellus's map is extremely difficult to read, even in the multispectral images, but two of those images do reveal some of the text (tile C04R04).[52] Given that the multispectral images reveal much more of the text on nearby cartouches, Martellus must have written the text on this cartouche using a different pigment. We see the same result in other parts of the map, particularly northeastern Asia, namely, that a given multispectral processing reveals some texts very clearly, but not others that are immediately adjacent, so Martellus used different pigments for several of the legends on the map. The legible words in this cartouche in the Indian Ocean are as follows:

[ ] qui et [ ]
[ ] habens [ ]rus pellem
duram et pilosam parit In terra et est
[ ]sus et iracundus nec de
[ ]tat locum ubi habitat

On its own this text is difficult to make sense of, but it was clearly the source of Waldseemüller's legend in the same location on his 1507 map:

> *Hic videtur felitus qui et vitulus marinus dicitur habens vocem ut thaurus pellem durissimam et vilosam parit in terra et est fortis et animosus et iracundus nec facile mutat locum ubi habitat*
>
> *Here is seen the felitus which is also called the sea cow; it has the voice of a bull and very hard, hairy skin. It gives birth on land, is strong and bold and irascible, and is reluctant to change where it lives.*

The text about this sea monster comes from the *Hortus sanitatis*.[53] So once again Martellus used the *Hortus Sanitatis* as a source for information about an animal, and Waldseemüller copied Martellus's legend onto his map.

The next cartouche to the south on Martellus's map is partly legible in one of the ultraviolet images of the map made in 1959 and much more readily legible in the multispectral images (tile C04R04 and C05R04) (Figs. 2.3 and 2.4). It reads:

Hic cernitur orca monstrum marinum
ad modum solis cum reuerberat cuius figura
vix describi potest nisi quod est pelle mollis et

---

[51] Johann Schöner on his globe of 1515 has a legend in this same part of the Indian Ocean that is very similar to Waldseemüller's and also an image of the leviathan that is copied from that in the *Hortus Sanitatis*—I compare Schöner's image with that in the *Hortus Sanitatis* in my *Sea Monsters on Medieval and Renaissance Maps* (London: British Library, 2013), p. 72. So Schöner perceived a connection between the legend and the *Hortus Sanitatis*.

[52] The two multispectral images in question are Supplementary Image 2.56 (C04R04b_bands01-22_RF+FL_cal_r90_cartouche_PC7) and Supplementary Image 2.57. (C04R04_RF+FL_cal_r90_med3_ratio_box101_stats_C06R02_PCA_R1G2B8_hue-45_sat-45).

[53] This information comes from one of two chapters in the *Hortus Sanitatis*, either "De piscibus," Chapter 38 on the "Felchus," or "De piscibus," Chapter 99 on the "Vacca et vitulus marinus."

Fig. 2.3 Natural light detail of the legend about the orca on the Yale Martellus map. (Courtesy of the Beinecke Rare Book and Manuscript Library)

Fig. 2.4 Multispectral image of the legend about the orca on the Yale Martellus map. (Image by Lazarus Project/ MegaVision/RIT/EMEL, courtesy of the Beinecke Rare Book and Manuscript Library)

> *carne in mensa* [sic].
>
> *Here is seen the orca, a sea monster that is like the sun when it shines, whose form can hardly be described, except that its skin is soft and its body huge.*

The corresponding legend on Waldseemüller's map is very similar indeed; it reads:

> *Hic cernitur orcha mirabile monstrum mari[n]um ad modum [s]olis cum reuerberat cuius figura vix describi potest nisi quod est pelle mollis et carne in mensa* [sic].
>
> *Here is seen the orca, an extraordinary sea monster, like the sun when it shines, whose form can hardly be described, except that its skin is soft and its body huge.*

Waldseemüller's borrowing from Martellus is perfectly clear here. The description of the orca comes from the *Hortus sanitatis*, "De piscibus," Chapter 64,[54] although that text does not contain the comparison with the sun.

The last of the five legends in the western Indian Ocean on Martellus's map is northeast of Madagascar. The text is legible in a natural light image of the map (in the multispectral images, it is on tile C06R04), and it reads:

> *Hic nascitur conchila maris qui murex*
> *dicitur que circumciso ferro lacrimas rubeas*
> *emittit ex qua fit purpura que ostium appellant.*

[54] The relevant part of the description of the orca in the *Hortus Sanitatis*, "De piscibus," Chapter 64, runs thus: *Orchum monstrum marinum est vt ait Plinius. eius imago nulla potest exprimi representatione: nisi quod sit mollis carnis immense....* See the chapter on the orca in Catherine Jacquemard, Brigitte Gauvin, and Marie-Agnès Lucas-Avenel, eds., *Hortus Sanitatis: Livre IV, Les poissons* (Caen: Presses Universitaires de Caen, 2013), available at http://www.unicaen.fr/puc/sources/depiscibus/consult/hortus_fr/FR.hs.4.64.

The legend on Waldseemüller's map is identical to Martellus's except for the omission of the "que" that precedes "circumciso" in Martellus's legend and the omission of the second "c" in "circumciso":

> *Hic nascitur conchila maris qui murex dicitur circum[c]iso ferro lacrimas rubeas emittit ex qua fit purpura que ostium appellant.*
>
> *Here is born the sea-shell called the murex, which, when it is cut with iron, sheds red tears from which is made the purple dye called* ostrum.

This description of the murex comes from the *Hortus sanitatis*, "De piscibus," Chapter 58[55]; the close similarity of Waldseemüller's legend to Martellus's leaves no doubt that Waldseemüller was borrowing from Martellus.[56]

On Martellus's map the island of Madagascar sits astride the Tropic of Capricorn. Even the name of the island, which is just north of the Tropic, is very difficult to read, but it is legible in the multispectral images (tile C05R04) as *Madargaschar insula*—an unusual spelling. Waldseemüller's spelling is more standard, *Madagascar*, so either he was consulting a different version of Martellus's map, or he modified what he found there. There are a few different descriptive texts on the island on Martellus's map; two of the texts to the north of the Tropic are illegible in natural light, infrared, and ultraviolet, but it is possible to read them in the multispectral images (tile C05R04):

> *Isti sunt macometisti*
> *habent copiam leonum elefan*
> *tum camelorum et ceterorum ani[malium]*
> *circuit tote Insule*
> *4000 passuum*
> *These people are Mahomedans; they have many lions, elephants, camels and other ani[mals].*
> *The circumference of the whole island is 4000 paces.*

The claim that the circumference of the island is 4000 paces is a mistake: the figure should be 4000 miles. This material comes from Marco Polo's description of Madagascar.[57]

[55] See the chapter on the murex in Catherine Jacquemard, Brigitte Gauvin, and Marie-Agnès Lucas-Avenel, eds., *Hortus sanitatis: Livre IV, Les poissons* (Caen: Presses Universitaires de Caen, 2013), available at http://www.unicaen.fr/puc/sources/depiscibus/consult/hortus_fr/FR.hs.4.58.

[56] The location of this legend near Madagascar would seem to derive from Ptolemy's *Geography* 7.2, where Ptolemy says that near the island of Bazakata, which he places in this part of the Indian Ocean, "some say there is found in abundance the murex," although there is no legend relating to the murex in this area on any Ptolemaic map that I have seen. For the passage in Ptolemy, see Alfred Stückelberger and Gerd Grasshoff, eds., *Klaudios Ptolemaios Handbuch der Geographie: griechisch-deutsch* (Basel: Schwabe Verlag, 2006), vol. 2, pp. 728–729; the English translation is from J. W. McCrindle, *Ancient India as Described by Ptolemy* (Calcutta: Thacker, Spink, & Co.; London: Trübner, 1885), p. 236.

[57] See Marco Polo, *Marka Pavlova z Benátek, Milion: Dle jediného rukopisu spolu s prilusnym zakladem latinskym*, ed. Justin Václav Prásek (Prague: Nákl. Ceské akademie císare Frantiska Iozefa, 1902), Book 3, Chapter 39, pp. 184–185; for Yule's translation see *The Book of Ser Marco Polo* (see note 44 in Chap. 2), Book 3, Chapter 33, vol. 2, pp. 411–413.

The texts in the southern part of the island are also very difficult to read. There is a short text near the island's southwestern coast of which only two words are legible even in the multispectral images,[58] and those are *silva sandali*, "forest of sandalwood"—and the same phrase appears in the northern part of the island on Waldseemüller's map. The following words are legible in the multispectral images (partly visible in tile C05R05, wholly visible in tile C06R05):

Hec insula habet [ ]
[ ] et
distat ab insula scoyra
[pe]r mille miliaria
[versus] austrum
This island has… it is 1000 miles distant from the island of Socotra to the south.

This material also comes from Marco Polo's chapter on Madagascar, which says that the island is 1000 miles south of Scoyra (Socotra).[59] On Waldseemüller's map Madagascar also sits astride the Tropic of Capricorn, and its longitudinal position is the same as on Martellus's map, centered at about 105° east: the island is closer to Africa on Martellus's map because of Martellus's large eastward-jutting peninsula in southern Africa. The identity in the positions of the island on the two maps is striking, yet there are differences in the legends on the island on the two maps. The northern of the two legends on Waldseemüller's map reads:

Silva sandali Hec est maior Et ditior insula totius mundi continet enim ambitus eius in circuitu miliaria 4000 habitatores sunt Saracen[ni] macometiste non habentes regem.

Forests of sandalwood. This is the largest and richest island in the whole world. For its circumference is 4000 miles, and its inhabitants are Islamic Saracens who do not have a king.

And the southern legend:

Habet haec insula nemora sandalorum & omnia genera specierum etiam elephantes leones linces leopardos ccrvos [sic, for cervos] damos et aves multarum specierum.

This island has groves of sandalwood and all types of spices, as well as elephants, lions, lynxes, leopards, deer, bucks, and many species of birds.

In a large cartouche just east of Madagascar, Waldseemüller does mention the distance from Moabar to Madagascar, whereas Martellus mentions the distance from Scoyra to Madagascar, and they use different units: Martellus speaks in terms of miles, while Waldseemüller refers to days of sailing.[60] Madagascar thus represents an interesting case in which Waldseemüller places the island in exactly the same location as Martellus, but does not closely follow Martellus's legends on the island. Either

[58] The image that best reveals these words is Supplementary Image 2.58 (C05R05_mada_RnoprocPC8_Gmnf4_B91bdPC12).

[59] For Marco Polo's text on Madagascar, see note 57 in Chap. 2.

[60] Waldseemüller's legend begins: *Veniunt de reguo* [for *regno*] *moabar naves in 20. diebus ad hanc insulam madagascar & fix iu* [for *in*] *tribus mensibus possunt redire in moabar....*

Waldseemüller was using a different version of Martellus's map than the one that has come down to us, or he was unsatisfied with the legends he found on the island on Martellus's map and consulted Marco Polo in order to create his own legends; or both of these factors influenced the legends on Waldseemüller's map.

Legends in the Eastern Indian Ocean

On Martellus's map there are seven banners with text in the eastern Indian Ocean between Taprobana to the west and the fourth peninsula to the east. In this part of the map, the correspondence between the locations of the legends on Martellus's and Waldseemüller's maps is not so striking, largely because Waldseemüller has more legends, and also because one of the legends that Martellus has just to the east of Taprobana, Waldseemüller has to the west of the island. Some of these legends come from Ptolemy and thus are of no evidentiary value as far as corroborating Waldseemüller's use of Martellus's map: Waldseemüller might have copied them from any number of manuscripts or printed editions of Ptolemy, rather than from Martellus.

Just to the east of Taprobana on Martellus's map, there is the following legend, which comes from Ptolemy 7.4, and thus (as just mentioned) has no evidentiary value to confirm Waldseemüller's reliance on Martellus. It is extremely difficult to read, even in the multispectral images (tile C06R04):

An[te taprobanam] cohortes sunt
Insularum quas dicunt esse .1378.
[ ]
[ ]

The corresponding legend on Waldseemüller's map is to the west of Taprobana and, in comparison with the passage about Taprobana in Ptolemy 7.4, omits the word *nomina* before *traduntur*:

Ante taprobanam chortes [sic] sunt insularum quas dicunt esse .1378. numero quarum tamen traduntur haec sunt

Near Taprobana there are many islands—they say there are 1378; these are the ones whose [names] are known.

South of the preceding legend on Martellus's map is the following one, which comes from Ptolemy 7.2 on the islands called the Maniolae, which are inhabited by man-eaters; magnetic rocks are found there which draw any ship with nails to its destruction.[61] The legend runs (tile C06R04):

[61] For discussion of these islands in Ptolemy, see W. J. van der Meulen, "Suvarnadvîpa and the Chrysê Chersonêsos," *Indonesia* 18 (1974), pp. 1–40, at 15; G. E. Gerini, *Researches on Ptolemy's Geography of Eastern Asia* (London: Royal Asiatic Society and Royal Geographical Society, 1909), pp. 420–424; and Richard Hennig, "Der Hafen Kattigara und der Magnetberg des Ptolemäus," *Klio* 23 (1929), pp. 256–276.

Maniole insule [ ] lapis
g[ ]gnitur herculeus obque hoc
navigia que clauos ferreos
habent detineri harum inco-
le antropofagi sunt.

The Maniolae Islands: [here] the rock of Hercules is produced, because of which ships with iron nails are detained; the inhabitants of these islands are cannibals.

Waldseemüller's legend is very similar but again could come directly from Ptolemy rather than from Martellus:

Hic lapis gignitur Herculeus obque hoc navigia que clauos ferreos habent detineri harum incole antropofagi sunt.

Here the rock of Hercules is produced, because of which ships with iron nails are detained; the inhabitants of these islands are cannibals.

The legend on Martellus's map that is northeast of the legend east of Taprobana is illegible in natural light, ultraviolet, and infrared images of the map but is just barely legible in the multispectral images (tile C07R04). It reads:

In hac Insula dicunt conchas multas
fieri et qui hanc incolunt nudi continue
degere dicuntur vocarique Gimnatas

In this island they say there are many shellfish; the inhabitants are said to be continuously nude and are called gymnatas.

This corresponds to Waldseemüller's legend about the island Basachata or Bazacata, which curiously is printed twice on Waldseemüller's map, and comes from Ptolemy 7.2. The northern of Waldseemüller's two legends about the island runs:

In hac dicunt conchas multas fieri et qui hanc incolunt nudi continue degere dicuntur vocarique gimnatas.

In this island they say there are many shellfish; the inhabitants are said to be continuously nude and are called gymnatas.

Just south of the preceding legend on Martellus's map is a legend about a sea monster, specifically about a fish called the *narcos* (the torpedo fish) that numbs the hand and the whole body of any fisherman who touches it. The spot on Waldseemüller's map where this legend might have been copied is occupied instead by his repetition of the legend about Bazacata. The legend on Martellus's map runs thus (tile C07R04):

Narco piscis tantae est uirtu[tis ut]
mediante lino et calamo ad ma-
num piscatoris accedit stupor et
insensibilitas & totum corpus stupe
facit nisi cito Hamum [dejiciat]

The torpedo fish has such power that through the fishing line and rod numbness and insensibility reach the hand of the fisherman and it makes the whole body numb unless he quickly [drops] the net.

This legend, like the ones about sea monsters in the western Indian Ocean discussed earlier, comes from the *Hortus sanitatis*, in this case "De piscibus," Chapter 63[62]:

> *Narcos: teste Aristotele: tante virtutis est piscis: ut mediante lino et calamo ad manum piscatoris hamum tenentis accedat stupor et insensibilitatis:immo et totum corpum stupefacit. nisi citius hamum deiiciat. Inde est quod etiam stuporifera ut papauer et huiusmodi cetera dicunt narcotica.*
>
> *The torpedo fish, according to Aristotle, has such power that through the fishing line and rod numbness and insensibility reach the hand of the fisherman who is holding the net, and indeed it makes his whole body numb unless he drops the net quickly. And it is from the name of this fish that soporifics such as poppies and others are called narcotics.*

The source of this material in the *Hortus Sanitatis* is not Aristotle but rather Alexander Neckam's *De naturis rerum*, Book 1, Chapter 44, probably written around 1180.[63] We can only speculate about why there is no similar legend on Waldseemüller's map: perhaps he was using a different version of Martellus's map that did not include the legend, or perhaps he saw it on Martellus's map but it did not interest him enough to include it on his map.

Martellus has a legend in the eastern Indian Ocean just west of the tip of Aurea Chersonesus, and just south of the equator, that is illegible in natural light, ultraviolet, and infrared images of the map but is barely legible in a few of the multispectral images (tile C07R04).[64] This legend relates to the Daruse Islands (from Ptolemy 7.2, where the name is *Barussae*), and it seems to read:

> *Daruse Insule in his antropofagi*
> *incole esse perhibentur*
> *The Daruse Islands: the inhabitants are held to be man-eaters.*

The corresponding legend on Waldseemüller's map is in the same location and is very similar (*daruse insule in his antorpophagi incole esse perhibentur*), but Waldseemüller might have taken the legend from Ptolemy rather than from Martellus.

A legend southeast of the preceding is also illegible in natural light, ultraviolet, and infrared images of the map but is legible in the multispectral images. It is about the island Ibadium (also from Ptolemy 7.2, where the name is *Iabadii insula*), and it reads (tile C07R04):

---

[62] See Catherine Jacquemard, Brigitte Gauvin, and Marie-Agnès Lucas-Avenel, eds., *Hortus Sanitatis: Livre IV, Les poissons* (Caen: Presses Universitaires de Caen, 2013), whose chapter on the *narco* is available at http://www.unicaen.fr/puc/sources/depiscibus/consult/hortus_fr/FR.hs.4.63.

[63] See Alexander Neckam, *Alexandri Neckam De naturis rerum libri duo* (London: Longman, Green, Longman, Roberts, and Green, 1863), p. 156: *De pisce qui narcos dicitur. Narcos piscis est tantae virtutis, ut dicit Aristoteles, quod mediante lino et calamo ad manum piscatoris calamum tenentis accedit stupor et insensibilitas. Immo et totum corpus obstupescet, nisi citius hamum dejiciat. Hinc est etiam quod stuporifera, ut papaver et hujusmodi, dicuntur narcotica. Sic sic, si vitium aliquod, maxime si luxuriam, tibi associare volueris, obdormiet et operatio per manum designata, et tota congeries operum, quae per corpus accipi solet.*

[64] One of the multispectral images in which the legend about the Daruse islands is relatively legible is Supplementary Image 2.59 (C07R04_RF+FL_cal_r90_med3_bands01-12_IndOc_cartouches_RPC04_GIC02_BIC12_hue-30).

hec insula est fertilissima
et aurum multum efficit
This island is very fertile and produces much gold.

The corresponding legend on Waldseemüller's map is in the same location but somewhat more elaborate (*Ibadium hoc est ordei insula fertilissima et aurum multum efficit*), so either Waldseemüller was using another version of Martellus's map as a source, or he supplemented what he found on Martellus's map with material from Ptolemy.

North of the preceding legend on Martellus's map, there is a legend about the islands of satyrs, which comes from Ptolemy 7.2.[65] Much of the legend is legible in natural light images of the map, but the multispectral images reveal all of the text (tile C07R04)[66]:

qui has insulas habitant
caudas habere dicuntur
quales satirorum pingunt
Those who inhabit these islands are said to have tails, like those they paint of the satyrs.

The corresponding legend on Waldseemüller's map is similar but more concise: *satirorum insule que caudas haberi dicuntur*, that is, "The islands of the satyrs, which are said to have tails."

Legends in the Proto-Pacific

In this area a number of the legends on the Yale Martellus map are based on Marco Polo. If it were merely the case that the Martellus and Waldseemüller maps contained information from Marco Polo, these legends would do little to corroborate Waldseemüller's reliance on Martellus, as Waldseemüller could have obtained the information directly from the work of the Venetian traveler. But in many cases, the information from Marco Polo appears in the same locations on the two maps, and in these cases its probative value for Waldseemüller's reliance on Martellus is high.

The disposition of the islands in the southern part of the Proto-Pacific, for want of a better name, superficially seems quite similar on the two maps, for on both there is a string of large islands astride the Tropic of Capricorn (Figs. 2.5 and 2.6). A look at Waldseemüller's map shows that these islands (from west to east Seylam, Peutam, Necura, Iava Maior, and Candin) come from Marco Polo, and the general similarity of their positions on the two maps makes it clear that Martellus influenced

[65] For discussion of the islands of the satyrs in Ptolemy, see van der Meulen, "Suvarnadvîpa and the Chrysê Chersonêsos" (see note 61 in Chap. 2) p. 20; and Gerini, *Researches on Ptolemy's Geography of Eastern Asia* (see note 61 in Chap. 2), pp. 707–724.

[66] The text of this legend does not appear in all of the multispectral images of this tile, but does appear clearly in Supplementary Image 2.60 (C07R04_RF FL_cal_r90_cartouches_bands01-14 18 16 21_ICA2_R2G3B6_boxb_subs50).

Fig. 2.5 Multispectral image of the islands of southeast Asia on the Yale Martellus map. (Image by Lazarus Project/MegaVision/RIT/EMEL, courtesy of the Beinecke Rare Book and Manuscript Library)

Fig. 2.6 The islands of southeast Asia on Waldseemüller's 1507 world map. (Courtesy of the Library of Congress)

Waldseemüller here. But in fact these islands represent some of the strongest evidence that Waldseemüller was using a different version of Martellus's large map than the exemplar at Yale. The positions of the central islands in the two groups are quite different, and yet the arrangement of the islands on Waldseemüller's map is very similar to that on the Martellus-Rosselli map (see Fig. 1.6). This indicates that Waldseemüller was using a version of Martellus's large map that was similar to the Yale map in that it included the large triangular peninsula jutting eastward from Asia, the island of Japan, and the numerous descriptive texts that Waldseemüller copied and yet had the islands in Southeast Asia arranged as they are on the Martellus-Rosselli map.

I examine the islands in Southeast Asia on the Martellus map from west to east. It is extremely difficult to read the text on all of these islands, even in the multispectral images. In the center of the westernmost of these islands on Martellus's map in the multispectral images, it is possible to make out *Seyla insula* (tile C07R04), written at a very steep angle, essentially parallel to the island's southeastern shore. And a few more words are legible along the island's southeastern shore: *ambitus... de 2,000 milia xl.* This same information appears in Waldseemüller's legend:

> *SEYLAM Hec insula est vna de maioribus et melioribus mundi habens In circuitu miliaria duo milia et .xl. insula hec habet regem ditissimum qui nulli tributarius est homines insule hu[ius] ydolatre sunt omnes nudi ambulant nullum bladum habent excepto riso habent lapides preci[osi].*
>
> *Zeilan [Ceylon]: This island is one of the largest and best of the world, having a circumference of 2040 miles. This island has a very rich king who is tributary to no one. The men of this island are all idolaters and go about nude; they have no grain except rice, and they have precious stones.*

This material comes from Marco Polo,[67] and as suggested above, the combination of similar text located on an island in the same location shows that Waldseemüller was borrowing from Martellus here rather than having taken the legend directly from Marco Polo. However, Waldseemüller's island of Seylam extends further to the southeast than that on Martellus's map at Yale and is more similar to the shape the island has on the Martellus-Rosselli map.[68] This is very good evidence that Waldseemüller was using a different version of Martellus's 280° image of the world than we have in the Yale Martellus map.

East of Seylam on the Yale map, there is a small island labeled *Angoma insula*, representing Angama (the Andaman Islands); on Waldseemüller's map there is no corresponding small island east of Seylam, and

[67] In the Naples manuscript of Marco Polo, the passage about Seylam/Ceylon is in Book 3, Chapter 22— see Marco Polo, *Marka Pavlova z Benátek, Milion* (see note 57 in Chap. 2), pp. 165–166; for Yule's translation, see *The Book of Ser Marco Polo* (see note 44 in Chap. 2), Book 3, Chapter 14, vol. 2, pp. 312–314.

[68] The legend on Seylam on the Martellus-Rosselli map reads *Hec insula habet in cirrcuiitum* [*sic*] *miliaria 2 M et est una de melioribus mundi*, "This island is 2000 miles in circumference and is one of the best in the world."

Waldseemüller applies the name *Angama* to a large island south of Necura. On Martellus's map there is a small island south of *Angoma insula*. In the northern part of the island in one of the multispectral images, it is possible to make out the name *Necur* [*A*], and in the southern part of the island there is text that reads (C08R04):

> *Etiam isti habent capita*
> *quasi canina*
> *These men also have heads like those of dogs.*

Here it begins to become clear that the depiction of these islands on the Yale Martellus map is confused, for Marco Polo says that the men with heads like those of dogs live on Angama, not on Necura,[69] and Waldseemüller correctly places this information on Angama:

> *isti sunt ydolatre bestialiter uiuunt omnium aromatum copia ibi est habent homines deformata capita quasi canina*
> *These men are idolaters who live like beasts. There is an abundance of all aromatics there. The men have deformed heads, like the heads of dogs.*

On Martellus's map, directly east of the island just mentioned is another; in the name of the island, it is only possible to read the letters *R…ON*,[70] and in the text further south on the island, it is just possible to make out the words *Circuitus totius* and two lines down, *capita canina*. I do not know what name the letters *R…ON* might be part of, so this island is quite mysterious, but the reference to the dog-like heads makes it seem that Martellus has placed information about Angama on three separate islands, a surprising result. This difference again suggests that Waldseemüller was consulting a different version of the Yale Martellus map than has come down to us (Angama does not appear on the Martellus-Rosselli map).

Returning north to the Tropic of Capricorn and continuing east on Martellus's map, there is a large island that we might expect—judging from Waldseemüller's map—to be Peutam, but the multispectral images reveal the name *IAVA* in the southern part of the island,[71] so this island is Java Minor. Waldseemüller has this island in a different position, not on the Tropic of Capricorn but at the southern edge of his map, and the Martellus-Rosselli map has it in the same location as Waldseemüller, at the southern edge of the map. In the Naples manuscript of Marco Polo, the text at the beginning of the chapter on Necuram is not at all clear what

[69] See Marco Polo, *Marka Pavlova z Benátek, Milion* (see note 57 in Chap. 2), Book 3, Chapter 21, p. 165; for Yule's translation, see *The Book of Ser Marco Polo* (see note 44 in Chap. 2), Book 3, Chapter 13, vol. 2, p. 309.

[70] The letters *R…ON* are visible in Supplemtary Image 2.61 (C08R04_RF+FL_cal_r90_med3_stats_C05R05_RF+FL_cal_r90_bands_05+06+12+16_lake_name_cyan_PCA_R1G2B2_hue-120).

[71] The name *IAVA* is most readily visible in Supplementary Image 2.62 (C08R04_RF+FL_cal_r90_med3-bands01-22_stats_from_C06R02_PCA_bands01-05+08-16+19+22_PCA_R1G2B4_hue180_2). There is text written almost vertically on the eastern coast of Java, but it is not legible, even in the multispectral images.

direction Necuram and Angama are from Java Minor,[72] and it seems that Martellus assumed that they were to the south, as he places them on his Yale map. At some point he reread the passage in Polo that emphasizes how far south Java Minor is[73] and decided that Java Minor must be to the south, as it is on the Martellus-Rosselli map. The difference in the locations of these islands between the Yale Martellus map and Waldseemüller's map is very strong evidence that Waldseemüller was consulting a version of Martellus's large world map different from the Yale map, on which the disposition of these islands was much more similar to what we see on the Martellus-Rosselli map.

The next island east on the Martellus map is *Pevtam*: the name can be seen in the multispectral images (tile C08R04) running across the middle of the island. A few letters can also be distinguished toward the northwestern shore of the island, but not enough to make sense of. And the next island east on Waldseemüller's map is *Pevtam* as well.

The next island on the Martellus map is no doubt Java Major: the name is not legible even in the multispectral images (tiles C10R04 and C10R03), but its size and orientation are excellent matches for Java Major on Waldseemüller's 1507 map. But on Waldseemüller's map, the next island east after Peutam is Necura, and south of Necura is Angama—and east of Necura is Java Major. So the order of these islands is quite different on the two maps. And again, the order of the islands on the Martellus-Rosselli map (Peutam, Necura, Java Major) is very similar to that on Waldseemüller's map, providing strong evidence that Waldseemüller was consulting a version of Martellus's large map similar to the Martellus-Rosselli map in this area.

There is good corrobating evidence for this different version of Martellus's large map in Martin Behaim's terrestrial globe of 1492. Behaim, who almost certainly used a map by Martellus as a source, has these islands arranged very much as they appear on the Martellus-Rosselli map and on Waldseemüller's 1507 map (Fig. 2.7).

Moving east along the Tropic of Capricorn on the Yale map, north of the Tropic, there are two small islands, the northern of which is labeled *sandur* and the southern *candur*. These two islands come from Marco Polo[74]; while on Martellus's map *sandur* is almost directly north of *candur*, on Waldseemüller's *sandur* is east of *candur*; on the other hand, the longitude of the islands is about the same on both maps, approximately 205° east.

[72] See Marco Polo, *Marka Pavlova z Benátek, Milion* (see note 57 in Chap. 2), p. 165.

[73] Marco Polo, *Marka Pavlova z Benátek, Milion* (see note 57 in Chap. 2), p. 160.

[74] For discussion of Candur and Sandur, see Silvestri, *De insulis*, pp. 146–147 and 500–501, respectively, in the edition by Montesdeoca (see note 44 in Chap. 2); Ivar Hallberg, *L'Extrême Orient dans la littérature et la cartographie de l'Occident des XIIIe, XIVe, et XVe siècles; étude sur l'histoire de la géographie* (Göteborg: W. Zachrissons boktryckeri a.-b., 1907), p. 162, s.v. "Condur," and p. 450, s.v. "Sandur,"; and Pelliot, *Notes on Marco Polo* (see note 44 in Chap. 2), vol. 1, pp. 404–407.

Fig. 2.7 The islands of Southeast Asia on Ravenstein's facsimile of Martin Behaim's globe of 1492 in Martin Behaim: His Life and His Globe. *(Courtesy of the David Rumsey Historical Map Collection)*

Further east on Martellus's map is an island that seems to correspond in terms of position and shape to *Candin* on Waldseemüller's map, though no words are visible on it on Martellus's map (tiles C10R03 and C10R04).

On Martellus's map between Java Major and the Asian mainland, there is a banner in which only the following words are visible, even in multispectral images of this tile (C09R04):

Hic videtur [ ]
[ ]
solum unum [ ]

There is a cartouche in exactly the same location on Waldseemüller's map, and the few words legible on Martellus's map match those on Waldseemüller's:

Hic videtur granus piscis
mire magnitudinis habens
solum unum oculum in fronte.
Here is seen the granus, a very large fish that has only one eye in its forehead.

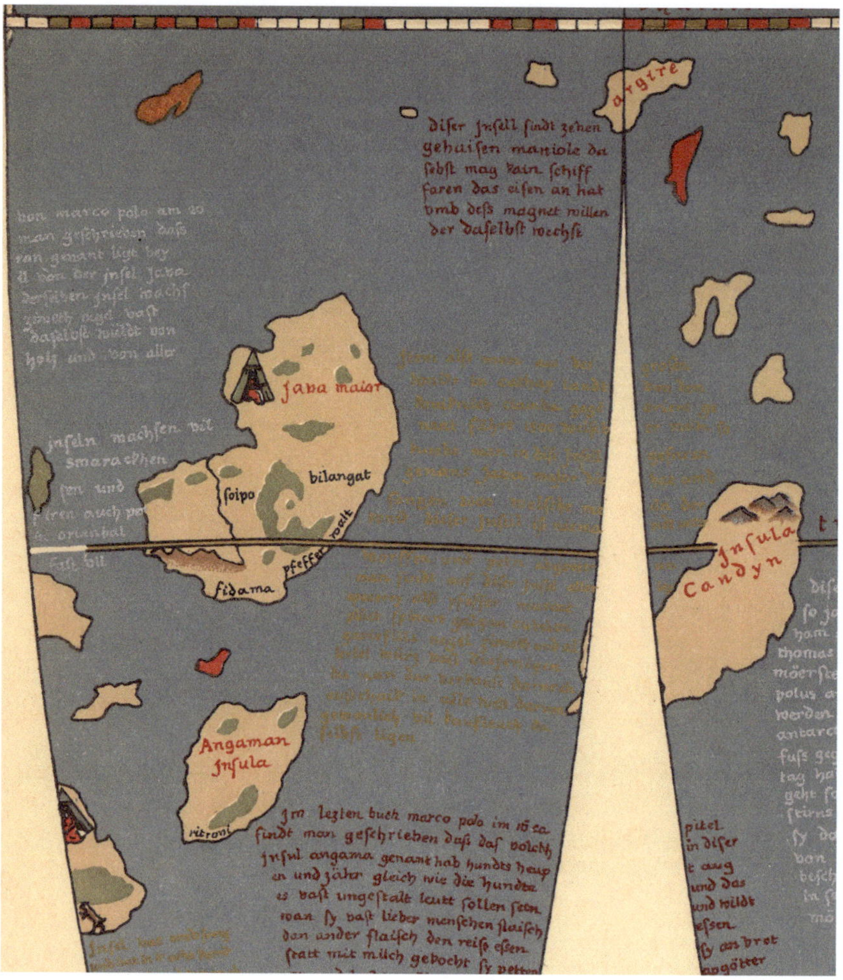

Fig. 2.7 (continued)

This sea creature appears in only three sources, namely, Thomas of Cantimpré,[75] Konrad von Megenberg,[76] and the *Hortus sanitatis*,[77] and given Martellus's use of the *Hortus Sanitatis* as a source for his other sea monsters, we may be confident that the same book was his source for this creature. And the correspondence of these two legends on the two maps is strong evidence that Waldseemüller was following Martellus.

[75]Thomas of Cantimpré, *Liber de natura rerum*, ed. H. Boese (Berlin and New York: De Gruyter, 1973), 7.40, p. 263

[76]See Conrad von Megenberg, *Das Buch der Natur*, ed. Franz Pfeiffer (Stuttgart: K. Aue, 1861), III.D.15, pp. 252–253.

[77]*Hortus sanitatis*, "De piscibus," Chapter 41. See the chapter on the *granus* in Catherine Jacquemard, Brigitte Gauvin, and Marie-Agnès Lucas-Avenel, eds., *Hortus sanitatis: Livre IV, Les poissons* (Caen: Presses Universitaires de Caen, 2013), available at http://www.unicaen.fr/puc/sources/depiscibus/consult/hortus_fr/FR.hs.4.41.

The four cartouches west of Japan are of particular interest as there are no corresponding cartouches on Waldseemüller's 1507 map. It is tempting to take this difference between the maps as additional evidence that Waldseemüller was working with a different version of Martellus's map than that at Yale, particularly as there was certainly room in the waters west of Japan on Waldseemüller's map for additional texts. There are no similar cartouches on Martin Behaim's globe of 1492 either, which may be taken as evidence that the version of Martellus's map that both Waldseemüller and Behaim consulted lacked these cartouches. These cartouches on the Yale Martellus map are exceedingly difficult to read. The northernmost cartouche is illegible in natural light, infrared, and ultraviolet images, but it is legible in some of the multispectral images (tile C10R02):

hic piscantur
perle rube
Here red pearls are fished up.

This text derives from Marco Polo, who mentions the red pearls of Japan.[78]

The next cartouche to the south is also very difficult to read, even in the multispectral images. It contains two lines, and the first line reads *hic piscantur coralli*, "Here they fish up coral," but I cannot make out the second line. In the next cartouche to the south on the Yale map, which is just south of the Tropic of Cancer (still tile C10R02), the first line seems to read *hic vedetur serena*, "here there is seen a siren," but I have not been able to read the second line. Waldseemüller has a cartouche about the siren well to the southwest, just south of the equator off the eastern coast of Asia[79]; this difference in the locations of the texts about sirens on the two maps is probably to be seen as evidence that Waldseemüller was using a different version of Martellus's large map than the one that has come down to us.

[78]In the Naples manuscript of Marco Polo, the passage about the pearls of Japan is at the end of Book 3, Chapter 2—see Marco Polo, *Marka Pavlova z Benátek, Milion* (see note 57 in Chap. 2), p. 154: *Ibi sunt margarite in copia maxima, que rotunde et grosse sunt rubeique coloris, que margaritas albas precio ac valore precellunt*. For Yule's translation, see *The Book of Ser Marco Polo* (see note 44 in Chap. 2), Book 3, Chapter 2, vol. 2, pp. 253–256. Incidentally there is a marginal annotation highlighting the red pearls in Christopher Columbus's copy of the c. 1484 edition of Marco Polo published by Leeu: see the facsimile of Christopher Columbus's copy of this book which has been published as *Libro de las maravillas del mundo: facsímil del que, usado por Cristóbal Colón, se encuentra depositado en la Biblioteca Capitular y Colombina del Cabildo Catedral de Sevilla* (Madrid: Testimonio, 1986).

[79]Waldseemüller's legend reads *hic videtur syrena horrible monstrum marinum*, "Here is seen the siren, a horrible sea monster."

The fourth of these cartouches is further south, just southwest of Japan. The text is very difficult to read, even in the multispectral images, but the legible words run (tile C10R02)[80]:

In omnibus [his insulis]
Reperiunt[ur] lapides precios[i]
Precious stones are found in all [of these islands].

This text seems to represent a general impression of the islands of the Indian Ocean from reading Marco Polo, rather than a quotation of a specific passage. West of this cartouche, there is a distinctive two-colored island that lacks a name or descriptive text; Behaim has an island labeled *crisis* in essentially the same location,[81] referring to the island of Chryse, an island in the Indian Ocean with golden soil that is mentioned by Pomponius Mela 3.7, Pliny 6.80, Solinus 52.17, and Isidore 14.6.11. Waldseemüller on his 1507 map has an island in much the same location, but without name or descriptive text. It is possible that on a different version of his large world map, Martellus labeled this island Chryse; it is also possible that Beheim was inspired to add the reference to Chryse by another map or text.

Martellus's depiction of Japan as oriented on a north-south axis, which has no basis in Marco Polo,[82] was very influential, much more influential than Fra Mauro's earlier depiction of the island in his *mappamundi* of c. 1450. Martellus's depiction influenced not only Behaim, Columbus, and Waldseemüller but also evidently Giovanni Matteo Contarini, who gives the island a north-south orientation on his 1506 world map (see Fig. 2.8), but could not have been influenced by Waldseemüller's later 1507 map; and also Rosselli, for the island has this same orientation in his oval planisphere of c. 1508.[83] Martellus's depiction of Japan also influenced, by way of Waldseemüller, Peter Apian's depictions of the island in his world maps of 1520 and 1530[84]; Benedetto Bordone's map of the island in his *Libro di Benedetto Bordone, nel qual si ragiona de tutte l'isole del mondo* (Venice: [N. d'Aristotile, detto Zoppino], 1528), Book 3, f. 68r; the depiction of Japan on Sebastian Münster's map of the New World titled *Novae Insulae, XVII Nova Tabula* first published in his edition of Ptolemy's *Geographia* (Basel:

[80] The text in the fourth of these cartouches is legible in Supplemenary Image 2.63 (C10R02_Japan_lowest_sel_ops123) and Supplemenary Image 2.64 (C10R02_Japan_pseudorender2).

[81] See Ravenstein, *Martin Behaim, His Life and His Globe* (see note 72 in Chap. 1), p. 89.

[82] For discussion of Polo's description of Japan, see K. Enoki, "Marco Polo and Japan," in *Oriente Poliano: studi e conferenze tenute all'Is. M.E.O. in occasione del VII centenario della nascita di Marco Polo (1254–1954)* (Rome: Istituto Poligrafico dello Stato, 1957), pp. 23–44, esp. 23–36.

[83] For an illustration and discussion of Rosselli's oval planisphere, see Shirley, *The Mapping of the World* (see note 54 in Chap. 1), p. 32; and Chet Van Duzer, "A Newly Discovered Fourth Exemplar of Francesco Rosselli's Oval Planisphere of c.1508," *Imago Mundi* 60.2 (2008), pp. 195–201.

[84] For illustrations and discussions of Apian's 1520 and 1530 world maps, see Shirley, *The Mapping of the World* (see note 54 in Chap. 1), pp. 51–53 and 68–69, respectively.

Fig. 2.8 World map by Giovanni Matteo Contarini, printed in 1506. (British Library Maps C.2.cc.4., © the British Library Board)

apud Henricum Petrum, 1540)[85]; and Sebastian Cabot on his world map of 1544.[86]

In orientation and outline, the depiction of Japan on the Yale Martellus map is very similar to the map of Japan in the Florence manuscript of Martellus's *Insularium* (Biblioteca Medicea Laurenziana, Pluteo 29.25, f. 76r) (see Fig. 1.9).[87] In that map the island bears its name, *ÇINPANGU INSULA*, "Cipangu Island"; a brief legend indicates that the island has *Silva nutum muscatarum*, "Grove of nutmeg"; a brief text beside an image of a city in the middle of the island reads *Cinpangu civitas regale*, "Cipangu the royal city"; and toward the southern end of the island, a legend reads *silva ebarii*, "forest of ebony." There are two longer legends on the island; the northern legend is a very abbreviated and hence confused account of the winds that blow near Japan and of the attempted Mongol invasion of the island in the thirteenth century[88]:

> *In hac insula duo tantum spirant venti naues provintie quinsay inde hic accedentes veniunt In yeme et in estate ad portes redeunt.*
>
> *In this island only two winds blow. Ships from the province of Quinsay come to Japan in the winter and return to [their] ports in summer.*

The source of the reference to the attempted invasion of Japan and to the winds is Marco Polo, of course.[89] The southern legend indicates the island's position:

> *Hec insula distat ab littora provintie mangi Centum et quingenta milia passum est magna valde*
>
> *This island is 100 and 500 miles from the shore of the Province of Mangi, and is very large.*

---

[85]Münster's map of the New World is conveniently reproduced in Nebenzahl, *Atlas of Columbus* (see note 42 in Chap. 1), pp. 98–99.

[86]There are two surviving copies of Cabot's map: Paris, Bibliothèque nationale de France, Rés. Ge AA 582; and Klassik Stiftung Weimar, Kt 020 - 31 S, which latter exemplar however lacks the surrounding printed text. The exemplar of Cabot's map in Paris is well reproduced in Nebenzahl, *Atlas of Columbus* (see note 42 in Chap. 1), pp. 106–107; on the exemplar discovered in Weimar, see Günter Schilder, *Monumenta cartographica Neerlandica* (Alphen aan den Rijn, Holland: Uitgevermaatschappij Canaletto, 1986-), vol. 2, p. 23. On Cabot and his map, see Karrow, *Mapmakers of the Sixteenth Century* (see note 5 in Chap. 1), pp. 103–112, esp. 108–109.

[87]The map of Japan in the Florence manuscript is illustrated in Roberto Almagià, "I mappamondi di Enrico Martello e alcuni concetti geografici di Cristoforo Columbo," *La Bibliofilia* 42 (1940), pp. 288–311, esp. p. 303, and Gentile, *Firenze e la scoperta dell'America* (see note 8 in Chap. 1), plate 46. For discussion see George Kish, "Two Fifteenth-Century Maps of 'Zipangu': Notes on the Early Cartography of Japan," *The Yale University Library Gazette* 40.4 (1966), pp. 206–214, who has a diagram of the outlines of Japan on Fra Mauro's map, Martellus's *Insularium*, the Yale Martellus map, and Behaim's globe on p. 211.

[88]The long legends on Japan in the Florence manuscript of Martellus's island book have been previously transcribed by Folker Reichert, "Zipangu. Marco Polos Japan und das europäische Weltbild zwischen Mittelalter und Neuzeit," in Sabine Klocke-Daffa, Jürgen Scheffler, and Gisela Wilbertz, eds., *Engelbert Kaempfer (1651–1716) und die kulturelle Begegnung zwischen Europa und Asien* (Lemgo: Landesverband Lippe, Institut für Lippische Landeskunde, 2003), pp. 147–168, at 155; and Folker Reichert, *Asien und Europa im Mittelalter: Studien zur Geschichte des Reisens* (Göttingen: Vandenhoeck & Ruprecht, 2014), p. 393.

[89]For Marco Polo's account, the Mongols attempt to invade Japan (conflating two attempts into one); in the Naples manuscript of his work, see Marco Polo, *Marka Pavlova z Benátek, Milion* (see note 56 in Chap. 2), Book 3, Chapters 3–6, pp. 154–156; the part about the winds comes from Book 3, Chapter 8, p. 158. For Yule's translation of these chapters see *The Book of Ser Marco Polo* (see note 43 in Chap. 2), Book 3, Chapters 2–3, vol. 2, pp. 253–260, and Chapter 4, vol. 2, pp. 23–265.

Martellus wrote *centum*, 100, by mistake for *mille*, 1000, giving the distance as 100 and 500 miles rather than 1000 and 500 miles, for Marco Polo gives the distance as 1500 miles (see just below).

The texts on Japan on the Yale Martellus map are extremely difficult to read. There is one text south of the Tropic of Cancer, immediately south of the label *Zinpanghu insula* and north of a chain of mountains that runs east and west across the island—a text that is totally invisible in natural light, infrared, ultraviolet, and indeed most multispectral images of the map but does appear in a few of the multispectral images (tile C10R02).[90] There are four lines of text, and only the first one is legible in these images:

> *Insula habundat auri*
> *The island abounds in gold.*

Marco Polo mentions Japan's abundant gold and the fact that the king does not permit it to be exported, and that as a result few merchants visit the island.[91] It is interesting that the material in this legend does not appear in either of the legends on Japan in the Florence manuscript of Martellus's *Insularium*.

The first few words of the southern legend on Japan on the Yale map are visible in natural light: Roberto Almagià read the first line as *Haec insula dist...*, and in the second line he thought he could make out the word *litora*.[92] By examining several different multispectral images of this part of the map, it is possible to read much more of the legend (tile C10R02) (Fig. 2.5):

> *Hec insula dist*
> *at ab continente pro*
> *vintie Mangi per*
> *mille miliaria habent*
> *linguam propriam et*
> *ambitus eius [ ]*
> *mille miliaria*
> *et in....*
> *This island is 1000 miles from the continent of the province of Mangi; the people have their own language and the circumference of the island is ??? miles and....*

[90]The two files that reveal the northern of the two texts on Japan are Supplementary Image 2.65 (C10R01_Japan_cal_r90_med3_stats_from_C06R02_PCA_R2G8B22_img2_Japan) and Supplementary Image 2.66 (C10R02_Japan_cal_r90_med3_22bands_C06R02_stats_PCA_R2G8B22).

[91]See Marco Polo, *Marka Pavlova z Benátek, Milion* (see note 56 in Chap. 2), Book 3, Chapter 2, p. 153; for Yule's translation, see *The Book of Ser Marco Polo* (see note 43 in Chap. 2), Book 3, Chapter 2, vol. 2, pp. 253–256.

[92]Almagià, "Worldmap by Henricus Martellus at Berne," June, 1960 (see note 2 in Front Matter), p. 4. Kish, "Two Fifteenth-Century Maps of 'Zipangu'" (see note 81 in Chap. 2), pp. 208–209, asserts that examination of the Yale Martellus map under various types of light confirmed that the legends on Japan on that map were very similar to those on Japan in the Florence manuscript of the *Insularium*, though he was only able to make out a few words on the Yale map. Reading the legends in the multispectral image contradicts this claim.

The first part of the legend is a good match for the corresponding one in the Florence manuscript of the *Insularium*, but strangely Martellus again makes an error with regard to the distance of the island from the mainland, asserting that the distance is 1000 miles instead of the 1500 miles reported by Marco Polo.[93] The remainder of the text on the Martellus map that is legible (there seem to be two and a half more lines beyond those that are legible) is very interesting, as it contains material that does not appear in the versions of Marco Polo that I have consulted. Polo does not specify that the Japanese have a distinct language, nor does he indicate the circumference of the island, and, in particular, this material does not appear in the Naples manuscript of Polo, which in other respects seems close to the version that Martellus was using. On the other hand, we might expect that Martellus would want to indicate a circumference for the island, as he indicates the circumference of almost every island in his *Insularium*.

Incidentally, the legend about the location of Cipangu in the Florence manuscript of Martellus's *Insularium* provides further evidence that Martellus was not using the printed edition of Marco Polo that was printed c. 1484 by Leeu. In that edition (Book 3, Chapter 2), the position of the island is described as follows: *Cyampagu que est insula ad orientem in alto mari distans a littore mangy per miliaria mille et cccc.*—that is, it indicates the distance as 1400 miles rather than 1500 miles. In the Florence manuscript of Martellus's *Insularium*, although he indicates the distance as 100 and 500 miles, this is evidently a mistake for 1000 and 500 miles. In the Naples manuscript of Marco Polo, the distance is given as 1500 miles.[94]

On Waldseemüller's map the island contains more indications of products available there than on the map in the Florence manuscript of Martellus's *Insularium*: in addition to the indication *nucum muscatum* (groves of nutmeg), there is *piper* (pepper), *zinamomum* (cinnamon), *gariofani* (cloves), *zinzib* (ginger), *spiconardi* (spikenard), and *silva sethim* (grove of sandalwood).[95] It seems that this same information does not appear on the Yale Martellus map: there are no signs of so many words near the edges of the island. It should be remarked that Marco Polo does not mention any of these specific spices in his description of the island; in fact while Polo alludes to there being many spices on the small islands

[93] The same distance from mainland Asia to Japan is indicated in the text on Japan in the 1522 edition of Ptolemy: *Hec insula distat versus orientem a litore magno miliaria mille sunt ydolatre habent regem nulli tributarium.*

[94] See Marco Polo, *Marka Pavlova z Benátek, Milion* (see note 56 in Chap. 2), Book 3, Chapter 2, p. 153.

[95] The identification of sethim or shittah with sandalwood is proposed by Jesús M.ª Porro Gutiérrez, "La cartografía ptolemaica del sureste asiático y su variante martelliana: planteamiento, consideraciones críticas y desarrollo de una hipótesis reinterpretativa," *Revista Complutense de Historia de América* 27 (2001), pp. 327–356, esp. p. 354.

around Japan,[96] he does not mention spices on Japan itself.[97] The longer legends on the island on Waldseemüller's map are as follows, first the northern one:

> *Hec insula distat versus orientum a littore magno per miliaria mille quingenta est magna valde habitatores sunt ydolatre habent regem nulli tributarium*
>
> *This island is 1500 miles east of the mainland; it is very large; its inhabitants are idolaters, and they have a king who is tributary to no one.*

And now the southern:

> *habent aurum in copia maxima sed non de facili eum extra insulam portare permittunt habent lapides preciosos de omni genere est ditissima valde super modum*
>
> *They have a great abundance of gold but they do not easily permit one to take it from the island. They have all types of precious stones and the island is extraordinarily rich.*

The material in these longer legends comes from Marco Polo.[98] The northern legend on Waldseemüller's map is similar to the southern one on the Yale Martellus map, but Waldseemüller makes no mention of the gold that Martellus speaks of in his northern legend. This difference, combined with the apparent lack of information about spices on either of Martellus's maps of Japan, suggests that either Waldseemüller made substantial revisions to what he found on Martellus's map, or he was consulting a different version of Martellus's map, or both.

It seems that Martin Behaim in making his globe of 1492 was using a version of Martellus's large world map that was more similar to that at Yale than the one that Waldseemüller was consulting. The fact that on Martellus's Yale map he addresses the circumference of Japan—even though it is not possible to read the number of miles Martellus specifies—is of great interest, for it tends to confirm that this map, or another very similar to it by Martellus, was the source of part of Behaim's legend on Japan, which says *voll hot umfang bei 1,200 meilen*, "Its compass is 1,200 miles."[99] In addition, we have seen that on his Yale map, Martellus refers to the presence of gold on Japan, while Waldseemüller does not, and Behaim also refers to the presence of gold on the island. He writes *Cipangu do wächßt vil gold*, "Cipangu where grows much gold," and he also mentions gold in two texts off the

[96] On the small islands around Japan, see Marco Polo, *Marka Pavlova z Benátek, Milion* (see note 56 in Chap. 2), Book 3, Chapter 8, pp. 157–158.

[97] On Japan itself see Marco Polo, *Marka Pavlova z Benátek, Milion* (see note 56 in Chap. 2), Book 3, Chapter 2, pp. 153–154; for Yule's translation, see *The Book of Ser Marco Polo* (see note 43 in Chap. 2), Book 3, Chapter 2, vol. 2, pp. 253–256.

[98] The text in the Naples manuscript of Marco Polo is supplied in Marco Polo, *Marka Pavlova z Benátek, Milion* (see note 57 in Chap. 2), Book 3, Chapter 2, pp. 153–154: *Çipangu insula est ad orientem in alto mari distans a littore mangy per milliaria mille quingenta et est magna valde. Habitatores eius albi sunt et stature decentis, ydolatre sunt et regem habent, sed nulli alii tributarii sunt. Ibi est aurum in copia maxima, sed rex non de facili eum extra insulam absportari permittit, propter quod mercatores pauci vadunt illuc et naues raro illuc ducuntur de regionibus aliis.... Ibi sunt margarite in copia maxima, que rotunde et grosse sunt rubeique coloris, que margaritas albas precio ac valore precellunt. Multi etiam sunt ibi lapides preciosi, propter quod insula çipangu opulentissima ualde est.* For Yule's translation, see *The Book of Ser Marco Polo* (see note 44 in Chap. 2), Book 3, Chapter 2, vol. 2, pp. 253–256.

[99] See Ravenstein, *Martin Behaim, His Life and His Globe* (see note 72 in Chap. 1), p. 89.

southern and eastern coasts of the island.[100] Thus the version of Martellus's map that Behaim was using was fairly similar to that at Yale; Waldseemüller was either using a modified version or was fairly aggressive in modifying the map by Martellus that he had before him or both.

LEGENDS ON LAND

Legends in Northern Asia

By far the greatest density of legends on the Yale Martellus map is found in northern Asia. I will go through the texts in northern Asia on the map from west to east, with comments on similarities with legends on Waldseemüller's 1507 world map. Legends that come directly from Marco Polo might seem to be of lesser value in terms of confirming Waldseemüller's use of Martellus, as Waldseemüller could have gathered the same information from his own study of Marco Polo. However, the coincidence of the locations of the same Polo-derived texts on the two maps provides strong evidence that Waldseemüller was using a map very similar indeed to Martellus's large world map at Yale. In some cases there are additional independent factors that corroborate Waldseemüller's reliance on Martellus for his legends.

Along the northern coast of Asia on the Yale Martellus map, there is a large bay between Scandinavia to the west and a large westward-jutting peninsula to the east. Along the western part of the shore of that bay, there is a lake drained by a river that feeds into the bay, and just north and northeast of that lake, the inscription *hiperborej scithe* is visible; on Waldseemüller's map the inscription *hiperborei scithe* is just south of the lake. The ethnonym is from Ptolemy, but neither the bay nor the lake has any basis in Ptolemy, which indicates that Waldseemüller was borrowing from Martellus here.

Just to the east, Martellus has a three-line legend on the coast, bits of which are legible in natural light, but more words appear in the multispectral images (tile C05R02). The legend begins with *hic*, and the second line clearly says *silvestres qui*, and Waldseemüller in the same spot has *hic viri silvestres reperiuntur*, "Here wild men are found." This legend comes from Marco Polo,[101] and the identity of the locations of the two legends confirms a very close relationship between the two maps.

---

[100] See again Ravenstein, *Martin Behaim, His Life and His Globe* (see note 72 in Chap. 1), p. 89.

[101] The passage about the wild men comes from Marco Polo, *Marka Pavlova z Benátek, Milion* (see note 57 in Chap. 2), Book 1, Chapter 62, p. 62: *Incole loci vocantur mecrith qui subiecti sunt magno kaam et habent tartarorum mores sunt autem silvestres homines*, with the phrase "sunt autem silvestres homines" coming from the early fifteenth-century manuscript Prague, Knihovna Pražské Metropolitní Kapituyí, G 28. For a description of this manuscript, see Consuelo Wager Dutschke, "Francesco Pipino and the Manuscripts of Marco Polo's Travels," Ph.D. Dissertation, University of California at Los Angeles, 1993, pp. 872–881. Yule's translation of this chapter is in *The Book of Ser Marco Polo* (see note 44 in Chap. 2), Book 1, Chapter 56, vol. 1, pp. 26–270.

Southeast of the preceding legend, there is text which is invisible in natural light but appears in some of the multispectral images (tile C05R02).[102] It reads:

hic magna habundantia
varorum armelinorum
[ ]rorum vulpium
cebelinorum
Here there is a great abundance of squirrels, ermines, [ ], foxes, and sables.

This legend comes from Marco Polo's chapter on the arms and clothes of the Tartars and agrees well with the text in the Naples manuscript of Polo.[103] There is no corresponding legend on Waldseemüller's map. The Martellus-Rosselli map has an abbreviated form of this legend, *Hic reperietur abundantia omnium varorum*, "Here there is found an abundance of all squirrels," in the same location as on the Yale map.

Northeast of the preceding legend on the Yale map, along the northern coast of Asia, in an area where no text is visible in natural, ultraviolet, or infrared light, the multispectral images reveal that there is indeed text (tile C06R02).[104] Even in these images, the text is very difficult to read; the legible parts are as follows:

[ ]
[ ]
[ ] jaciens spinas in vena
tores et canes habitat in cav
ernis circa mare

Waldseemüller has a closely related text in this same location that reads *hic histrix reperitur mire magnitudinis habitat in cavernis iaciens spinas in uenatores et canes*, that is, "Here is found the porcupine, which is very large; it lives in caves and throws its spines at hunters and dogs." The myth that the porcupine can hurl its quills goes back to Aristotle (*History of Animals* 9.26), and several authors, including Pliny (*Naturalis Historia* 8.53.125), mention that the porcupine hurls its quills at dogs and that it hibernates.[105] But the sources whose description is by far the most similar to Waldseemüller's legend are Thomas of Cantimpré[106] and the *Hortus*

[102]See, for example, Supplementary Image 2.67 (C05R02_bands01-22_RF+FL_cal_r90_med3_stats_C06R02_PCA_R14_stats_C03R02_G7B7) and 2.68 (C05R02_bands01-22_RF+FL_cal_r90_med3_difficult_cartouche_stats_C03R03_PC7).

[103]See Marco Polo, *Marka Pavlova z Benátek, Milion* (see note 57 in Chap. 2), Book 1, Chapter 56, pp. 56–57. For Yule's translation of the similar material in his edition, see *The Book of Ser Marco Polo* (see note 44 in Chap. 2), Book 1, Chapter 53, vol. 1, p. 257.

[104]Two of the best images for trying to read this legend are Supplementary Image 2.69 (C06R02_composite_hueangle) and Supplementary Image 2.70 (C05R02_bands01-22_RF+FL_cal_r90_med3_ratio_box71_cartouche_stats_PCA_R1G2B12_hue-180_G).

[105]Some discussion of the myths surrounding the porcupine may be found in Dieter Bitterli, "Exeter Book Riddle 15: Some Points for the Porcupine," *Anglia: Zeitschrift für englische Philologie* 120.4 (2002), pp. 461–487.

[106]Thomas of Cantimpré, *Liber de natura rerum*, ed. H. Boese (Berlin and New York: De Gruyter, 1973), 4.52, p. 138.

sanitatis.[107] Both mention that the creature lives near the sea, which is where Martellus and Waldseemüller locate their legends; both mention that the porcupine lives in caves, without referring to hibernation; and both mention that it hurls its quills at the dogs and men who hunt it. These sources are the only two I know that mention that the porcupine lives by the sea, and also the only two that specify that the porcupine lives in caves rather than burrows. Given Martellus's heavy use of the *Hortus Sanitatis* elsewhere, it is very likely that that book was the source of his legend about the porcupine and essentially certain that Waldseemüller took his legend from a map by Martellus.

Moving further east on the Yale map, we come to the large northwestward-jutting peninsula referred to earlier. This peninsula would seem to be an invention of Martellus: the only earlier maps on which I have seen it are the world maps in manuscripts of Martellus's *Insularium* and the Martellus-Rosselli map. So the appearance of this peninsula on Waldseemüller's map is strong evidence that he was borrowing from Martellus. Judging from the legend that Martellus places on this peninsula, it was his location for Marco Polo's Land of Darkness,[108] which Polo places in the far north—*in extremis habitacionibus septentrionalibus.*

Much of Martellus's legend on the peninsula is legible in natural light, but the last few words are only visible in multispectral images (tile C06R02). The legend runs:

> *harum partium Incole carent*
> *Rege bestialiter vivunt sunt magni fortes*
> *pulchres sed absque ingenio.*
> *The inhabitants of this area lack a king and live like beasts; they are large and strong and handsome but without intelligence.*

The legend comes from Marco Polo's description of the Land of Darkness[109]:

> *Sunt autem homines illius regionis pulcri magni et corpulenti sed sunt palidi valde. Regem non habent neque principem cuius sint dicioni subiecti sed incultorum morum homines sunt bestialiterque viventes.*
> *But the men of that region are handsome, large, and corpulent, but very pale. They have neither king nor ruler, but they are men of rude habits and they live like beasts.*

Given this text in Marco Polo, it would certainly seem that the word following *sed* in Martellus's legend should be *pallidi*, but I do not see it there. Waldseemüller's legend on this same peninsula runs:

---

[107] *Hortus Sanitatis*, "De animalibus," Chapter 73, "Histrix".

[108] Polo's chapter on the Land of Darkness in the Naples manuscript is in Book 3, Chapter 49, see Marco Polo, *Marka Pavlova z Benátek, Milion* (see note 57 in Chap. 2), pp. 197–198; for an English translation, see Yule, *The Book of Ser Marco Polo* (see note 44 in Chap. 2), Book 4, Chapter 21, vol. 2, pp. 484–485. For discussion of the region, see Pelliot, *Notes on Marco Polo* (see note 44 in Chap. 2), vol. 2, pp. 616–624.

[109] For the text on the Region of Darkness in the Naples manuscript of Polo, see Marco Polo, *Marka Pavlova z Benátek, Milion* (see note 57 in Chap. 2), Book 3, Chapter 49, p. 197. The English translation here is my own; Yule's version of Polo's text is quite different in *The Book of Ser Marco Polo* (see note 44 in Chap. 2), Book 4, Chapter 21, vol. 2, p. 485.

Incole harum carent rege proprio bestialit[er] vivunt sunt magni fortes et pulcri homines carent vino et blada sed divites carnium

The inhabitants of these [regions] do not have their own king, live like beasts and are big, strong and handsome men; they have neither wine nor corn but are rich in meat.

The first part of Waldseemüller's legend is very similar to Martellus's, but the second part comes from a different chapter in Marco Polo that describes another region in the far north, the plain north of Caracoron, about which Polo writes[110]:

Carnibus vescuntur animalium, que in venacionibus capiunt, et specialiter cervorum, de quibus copiam habent, quos eciam domesticant et factos domesticos equitant. Blado carent et vino.

They eat the flesh of animals which they catch by hunting, particularly of deer, of which they have an abundance, and in fact they domesticate deer and then ride them. They lack corn and wine.

Waldseemüller's combination of text pertaining to two disparate regions and two different chapters of Marco Polo is interesting; as it turns out, he did not have to look far on Martellus's map for the information about the inhabitants of the plain north of Caracoron: at the southern base of the eastward-jutting peninsula is the text *isti carent vino et blada*, "These men do not have wine or corn."

Just to the south is a text of which just a few letters are visible in natural light images, but the text is legible in the multispectral images, albeit with some difficulty (tile C06R02):

ab hinc portantur pelles [fe]rar[um]
ad partes cristianorum occidentales
From here the pelts [of wild beasts] are carried to Christian regions in the West.

This legend seems to have been the source of the similar legend in the same location on Waldseemüller's map:

hinc portantur pelles ferarum ad partes occidentales christianorum et qui habitant has regiones que longa est .10. dietarum habent regem de stirpe magni cham sunt idolatre et aliqui adorant natigas.

From here the pelts of wild beasts are carried to Christian regions in the West, and those who live in this region, which is a ten days' journey long, have a king who descends from the Great Khan. They are idolaters, and some worship Natigas.

Natigas is a Mongolian divinity.[111] This legend is interesting. Marco Polo does mention the abundance of furs in this region, but does not specify that they are taken to the West, so perhaps the cartographer was elaborating on his source here.[112] The part about the region being a 10-day journey long is significant as it demonstrates very clearly that Waldseemüller

[110]The text is from Book 1, Chapter 62, in the Naples manuscript of Polo: see Marco Polo, *Marka Pavlova z Benátek, Milion* (see note 57 in Chap. 2), p. 62. The English translation here is my own, but Yule's can be consulted in *The Book of Ser Marco Polo* (see note 44 in Chap. 2), Book 1, Chapter 56, vol. 1, p. 269.

[111]For discussion of Natigas, see Pelliot, *Notes on Marco Polo* (see note 44 in Chap. 2), vol. 2, pp. 791–792, s.v. "Natigai".

[112]For the text in the Naples manuscript, see Marco Polo, *Marka Pavlova z Benátek, Milion* (see note 57 in Chap. 2), Book 3, Chapter 47, pp. 194–195. For Yule's translation of this chapter, see *The Book of Ser Marco Polo* (see note 44 in Chap. 2), Book 4, Chapter 20, vol. 2, pp. 479–481.

and probably his source here, Martellus, were not using the printed edition of Marco Polo of c. 1484: that edition indicates that the length of the region is 40 days (Book 1, Chapter 62), while the Naples manuscript indicates a length of 10 days, as Waldseemüller does.[113]

There is a similar legend in almost the same location on the Martellus-Rosselli map, *ab hinc portantur pelles ferarum a partes occidentalis*, "From here the pelts of wild beasts are carried to regions in the West." The fact that such a similar legend appears in almost the same location on another map by Martellus confirms that Waldseemüller copied this legend from a map by Martellus.

Further south on the Yale Martellus map, there is a legend that Waldseemüller incorporated into the second half of his legend just cited. The text on the Martellus map is invisible in natural light, but much of it is legible in some of the multispectral images (tile C06R02). It runs:

qui habitant has regiones que est
longa 10 dierum iter habent regem
de stirpe magni cham adorant natigas

Those who live in this region, which is a ten day's journey long, have a king who descends from the Great Khan. They worship Natigas.

Further south on Martellus's map, between two branches of the Imaus Mountains, there is a legend which is invisible in natural light (see Fig. 2.9) but legible in some of the multispectral images. It reads (tile C06R02) (Fig. 2.10):

hic parviosa monstra
sunt similia hominibus
*tam dasa [*sic*] auri*
um magnitud
ine ut totum
corpus
tegunt

Here there are "parviosa," monsters similar to humans whose ears are so large that they can cover their whole body [with their ears].

These creatures are *panotii*, one of the monstrous peoples described by Pliny and other authors.[114] The legend does not come from Ptolemy or Marco Polo; its source may be Isidore's *Etymologiae*, for Isidore's phrasing

[113] For the text about the ten-day length of the province, see Marco Polo, *Marka Pavlova z Benátek, Milion* (see note 57 in Chap. 2), Book 1, Chapter 62, p. 62. Yule in his version of Polo does not include the ten-day figure: see *The Book of Ser Marco Polo* (see note 44 in Chap. 2), Book 1, Chapter 56, vol. 1, pp. 269–270.

[114] On the *panotii* see Hallberg, *L'Extrême Orient* (see note 74 in Chap. 2), pp. 391–392, which includes a list of variant spellings of the name, and Claude Lecouteux, "Les Panotéens: sources, diffusion, emploi," *Études germaniques* 35 (1980), pp. 253–266. For general discussion of the monstrous peoples, see John B. Friedman, *The Monstrous Races in Medieval Art and Thought* (Cambridge, MA, and London: Harvard University Press, 1981; reprinted with expanded bibliography Syracuse: Syracuse University Press, 2000).

Fig. 2.9 Natural light image of the legend about the panotii *on the Yale Martellus Map (courtesy of the Beinecke Rare Book and Manuscript Library)*

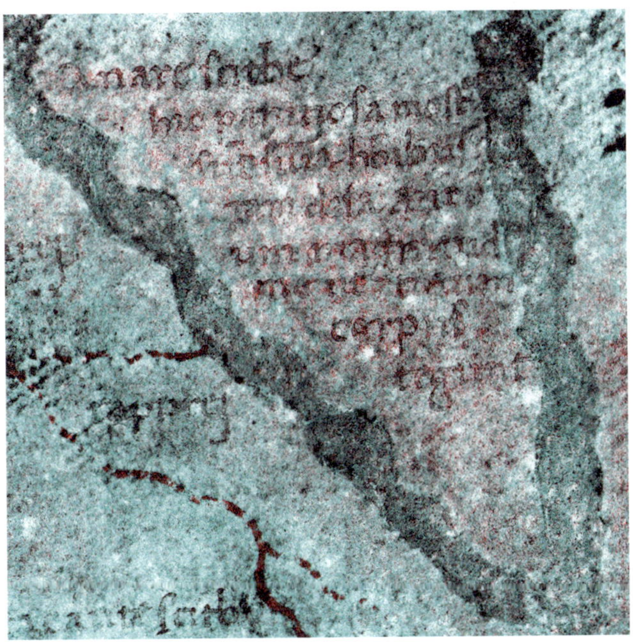

Fig. 2.10 Multispectral image of the legend about the panotii *(tile C06R02) on the Yale Martellus map. (Image by Lazarus Project/MegaVision/RIT/EMEL, courtesy of the Beinecke Rare Book and Manuscript Library)*

in his passage about the *panotii* is similar to that of Martellus.[115] There are two difficult words in this legend. The first is what is either the name or a descriptive adjective for the monsters, *paruiosa*, which I have never seen

[115]Isidore 11.3.19, *Panotios apud Scythiam esse ferunt, tam diffusa aurium magnitudine, ut omne corpus ex eis contegant.*

used for *panotii*. It might be taken as meaning "small-faced," but that meaning is not really apt for *panotii*. The second is *dasa* in the third line, which makes no sense, and to judge from Waldseemüller's corresponding legend (transcribed momentarily) seems to be an error for *diffusa*. There is a very similar legend in the same location—also between two branches of the Imaus Mountains—on Waldseemüller's map:

> *Monstra similia hominibus tam difusa aurum magnitudine ut totum corpus tegant hic sunt*
>
> *Here there are monsters similar to humans whose ears are so large that they can cover their whole body [with their ears].*

The questionable words in Martellus's legend were perhaps the result of mistakes in copying the text. Waldseemüller may have been working from a different version of Martellus's map that had *diffusa* rather than *dasa*, or else Waldseemüller realized that the source of the legend was Isidore 11.3.19, who uses *diffusa*, and corrected *dasa* to *difusa* himself.

Just east of the preceding legend on the Yale map, in an area ringed by trees, there is a legend that is invisible in natural light, infrared, and ultraviolet images but is visible in one of the multispectral images. Even in that image it is very faint, but it reads (tile C06R02):

> *Ippopedes hic reperiuntur*
>
> *habent humanam formam*
>
> *sed equinos pedes*
>
> *Here are found the Hippopodes: they have a human form but the feet of horses.*

The *hippopodes* are another of the famous monstrous peoples of Pliny.[116] Ptolemy in his *Geography* 3.5 mentions Hippopodes in Europe, not in Asia, and Marco Polo does not mention them at all; the source of this legend, like that on the *panotii*, is probably Isidore's *Etymologiae*.[117] Waldseemüller has a very similar legend in a similar location on his 1507 map:

> *hippedes hic reperiuntur habent*
>
> *bumanam* [sic] *formam sed equinos pedes.*
>
> *Here are found the Hippopodes: they have a human form but the feet of horses.*

Waldseemüller was clearly following Martellus here in terms of text, but while Martellus places this monstrous people within an enclosure of trees, Waldseemüller has them in an open plain. On the Martellus-Rosselli map, there is a very similar legend (*hippopedes hic reperiuntur habet humanam forma* [*sic*] *sed equinos pedes*), but there is no enclosure around the text.

[116] On the *hippopodes* see Pliny 4.13.95; Hallberg, *L'Extrême Orient* (see note 74 in Chap. 2), pp. 240–241, and Pentti Aalto and Tuomo Pekkanen, *Latin Sources on North-Eastern Eurasia* (Wiesbaden: Harrassowitz, 1975–1980), vol. 1, pp. 197–198.

[117] The phrasing of Martellus's legend is similar to that in Isidore's *Etymologiae* 11.3.25: *Hippopodes in Scythia sunt, humanam formam et equinos pedes habentes.*

This difference in the way the *hippopodes* are situated may indicate that Waldseemüller was working from a different version of Martellus's large world map—in which this detail was more similar to the arrangement on the Martellus-Rosselli map—or he may simply have chosen to modify what he found on Martellus's map. On his *Carta marina* of 1516, Waldseemüller enclosed the various monstrous peoples with mountain ranges.[118]

Moving northeast to the northern coast of Asia on the Yale map, we come to two lakes from each of which multiple rivers flow. Both of the lakes are named; in both cases the word *lacus* is legible; I cannot make out the names. But neither bears the name *nubius lacus*, which is the name of the lake in the same area on Waldseemüller's map. This difference between the maps is interesting. There is a lake very similar to Waldseemüller's lake in Martellus's world map in the London manuscript of his *Insularium*, where it bears the name *merardus lacus*; there are also similar lakes on the maps in the Florence (Fig. 1.2) and Leiden (Fig. 1.3) manuscripts of his *Insularium*, though the similarity is less striking and the lakes are somewhat further to the south. The difference between the Yale Martellus map and Waldseemüller's map in this regard may perhaps be taken as additional evidence that Waldseemüller was consulting a different version of Martellus's map.

Just southeast of the eastern lake on Martellus's map is the following legend, which is legible in natural light and also in the multispectral images (tile C07R02):

Hic antropo
fagi qui come
[dunt] carnes
humanas sunt
pessime nature
Here there are cannibals who eat human flesh—they are of the worst character.

In the same area, Waldseemüller has a legend that reads *Antropophagi edunt carnes humanas*, "Cannibals [who] eat human flesh." The source of this legend would seem to be Ptolemy 6.17, and indeed Martellus locates *Antrophagorum genus* in this same part of Asia in the world map in the Magliabechiano manuscript of Ptolemy's *Geography* that he made. In this case, the fact that Waldseemüller's text is longer than Ptolemy's and similar to Martellus's suggests that he was copying from Martellus.

A short distance south of this area on Martellus's map is a legend that is invisible in natural light and only legible in the multispectral images (tile C07R02). It pertains to the province of Chinchintalas and reads:

In hac provincia
sunt ydolatre et ali
qui cristiani eciam
macometisti sunt

[118] Chet Van Duzer, "A Northern Refuge of the Monstrous Races: Asia on Waldseemüller's 1516 *Carta marina*," *Imago Mundi* 62.2 (2010), pp. 221–231.

subiecti et pertin

ent ad magnam pro

vinciam tangut

In this province there are idolaters and some Christians, even Mohammedans. They are subjects to and belong to the great province of Tangut.

Waldseemüller on his 1507 map has a similar legend in the same area, just south of *Camul civitas*:

*hac [*sic*] provincia iacet inter duo deserta sunt ydolatre tamen eciam aliqui christiani obediunt magno cham sub dominio provincie tangut*

This province lies between two deserts. The people are idolaters, however there are some Christians. They obey the Great Khan, and are under the power of the province of Tangut.

The legend comes from Marco Polo, and in this case it is clear that Waldseemüller was consulting a different version of Martellus's large world map than the one that has come down to us, or else he checked the text of Marco Polo and modified what he found on Martellus accordingly: Martellus does not mention the Great Khan here, but Waldseemüller does, and Marco Polo does as well.[119]

Further east along the northern coast of Asia on Martellus's map is the following legend, which is legible both in natural light and in the multispectral images (C07R02)[120]:

hec regio vocatur balor [cuius] inco

le habitant [in montib]us sunt homines

silvestres [carent] vino et

blada utuntur carnibus cer

uorum et equitant cer

uos domesticos

This region is called Balor; its inhabitants live in the mountains, are wild men, and lack wine and corn. They eat the flesh of deer and ride domesticated deer.

Waldseemüller's legend is almost identical:

Balor regio Incole istius regionis incolae habitant in montibus sunt homines silvestres carent vino et blada utuntur carnibus cervorum et equitant cervos domesticos.

The region of Balor: the inhabitants of this region live in the mountains, are wild men, and lack wine and corn. They eat the flesh of deer and ride domesticated deer.

The information in the legend comes from Marco Polo's chapter on Balor[121]; yet we can be quite certain that Waldseemüller was using Martellus

[119] See Marco Polo, *Marka Pavlova z Benátek, Milion* (see note 57 in Chap. 2), Book 1, Chapter 47, p. 48. For Yule's translation see *The Book of Ser Marco Polo* (see note 44 in Chap. 2), Book 1, Chapter 42, vol. 1, pp. 212–213.

[120] There is a similar albeit shorter legend in this same region on the Martellus-Rosselli map: *Balor regio in regionibus istis habitant homines silvestres equitant cervos domesticos.*

[121] The readings in Naples, Biblioteca Nazionale Vittorio Emanuele III, Vind. lat. 50, are *burgi* and *bangu*; see Marco Polo, *Marka Pavlova z Benátek, Milion* (see note 57 in Chap. 2), Book 1, Chapter 62, p. 62, while in the edition of c. 1481 (still Book 1, Chapter 62), it is *Burgi* and *Burgy*. For Yule's translation of this chapter and notes on it, see *The Book of Ser Marco Polo* (see note 44 in Chap. 2), Book 1, Chapter 32, vol. 1, pp. 172 and 178–179. There is discussion of Balor in Hallberg, *L'Extrême Orient* (see note 74 in Chap. 2), p. 74, s.v. "Belor", and Henry Yule, "Notes Regarding Bolor, and Some Other Names in the Apocryphal Geography of the Upper Oxus," *Journal of the Royal Geographical Society* 42 (1872), pp. 473–480, esp. 473–475.

here: the phrasing of their legends is very similar, as is the location, and moreover both refer to the region as Balor, rather than as Bargu, which is the more usual spelling.

Directly south of the preceding legend on Martellus's map and just above the toponym *Chinchitila Provintia*, there is the following legend, which is legible in natural light and in the multispectral images (tile C07R02):

longitudo iste provincie
16 dierum sunt cristi
ani nestoriani
The width of this province is [a journey of] sixteen days; there are Nestorian Christians.

This legend comes from Marco Polo's chapter on Chinchintalas,[122] but Waldseemüller chose not to copy the legend—if it was on the version of Martellus's map that he saw. But it is certainly possible that Waldseemüller omited what he saw as less important legends that he found on Martellus's map.[123]

Just to the east on the Yale map is a region entirely enclosed by mountains and within it a legend that reads (tile C07R02 or C08R02):

Hic Judei clausi per
regem alexandrum
Here are the Jews enclosed by King Alexander [the Great].

The idea that Alexander enclosed the ten exiled tribes of the Jews as punishment for their apostasy from the true God is a variant of the story that Alexander imprisoned the evil races of Gog and Magog,[124] who (it was believed) would burst forth in the last days to ravage the earth (Ezekiel 38:1–39:16; Revelation 20:7–10).[125] The location of the enclosed nations in

[122] For the text in the Naples manuscript, see *Marka Pavlova z Benátek, Milion* (see note 57 in Chap. 2), Book 1, Chapter 47, p. 48; for Yule's translation, see *The Book of Ser Marco Polo* (see note 44 in Chap. 2), Book 1, Chapter 42, vol. 1, pp. 212–213.

[123] Waldseemüller does include information about the width of Chinchintalas on his *Carta marina* of 1516, but given that his legend on this map is not particularly similar to Martellus's, it seems likely that it was the result of independent consultation of Marco Polo. The legend runs *Chingitalis Provincia Hec provincia longa est .16. dierarum habens multas civitates et castra gentes sunt ydolatre et aliqui macomethani aliqui cristiani habente .3. ecclesias nestorianas. ibi eciam fit Salamandra.*

[124] The literature on Gog and Magog is substantial: in addition to the article by Gow just cited, see Andrew Runni Anderson, *Alexander's Gate, Gog and Magog, and the Inclosed Nations* (Cambridge, MA: The Medieval Academy of America, 1932), esp. pp. 14 and 70–86 on identifications of the tribes of Israel with Gog and Magog; Raoul Manselli, "I popoli immaginari: Gog e Magog," in *Popoli e paesi nella cultura altomedievale: settimane di studio del Centro Italiano di Studi sull'Alto Medioevo, Spoleto, 23–29 Aprile 1981* (Spoleto: Centro Italiano di Studi sull'Alto Medioevo, 1983), vol. 2, pp. 487–521; Danielle Lecoq, "L'image d'Alexandre à travers les mappemondes médievales (XIIe–XIIIe)," *Geographia Antiqua* 2 (1993), pp. 63–103, esp. pp. 92–96; Scott D. Westrem, "Against Gog and Magog," in Sylvia Tomasch and Sealy Gilles, eds., *Text and Territory: Geographical Imagination in the European Middle Ages* (Philadelphia: University of Pennsylvania Press, 1998), pp. 54–75; Andrew Gow, "Gog and Magog on *Mappaemundi* and Early Printed World Maps: Orientalizing Ethnography in the Apocalyptic Tradition," *Journal of Early Modern History* 2.1 (1998), pp. 61–88; and Hallberg, *L'Extrême Orient* (see note 74 in Chap. 2), pp. 225–230, s.v. "Gog et Magog".

[125] On the relationship between the stories of Gog and Magog on the one hand and the ten exiled tribes of the Jews on the other, see D. J. A. Ross, *Alexander historiatus: A Guide to Medieval Illustrated Alexander Literature* (Frankfurt am Main: Athenäum, 1988), pp. 34–35; the version involving the ten tribes was apparently the invention of Petrus Comestor in his *Historia Scholastica* Lib. Esther 5.

the far northeast is the result of the conflation of their story with that of Gog and Magog, for Marco Polo in his chapter on Georgiana places the mountains where Alexander enclosed the races near the Caspian, but in his chapter on the Province of Tenduc, he places Gog and Magog in the far northeast.[126] And these are the places where they are located on earlier maps; for example, on the Borgia XVI metal *mappamundi* of the first half of the fifteenth century in the Biblioteca Apostolica Vaticana, the *porte ferre* or iron gates with which Alexander was supposed to have enclosed these peoples are located west of the Caspian Sea, while the legends relating to the enclosure of Gog and Magog are behind mountains in the far northeastern part of the world.[127]

The location of the *judaei clausi* on the Yale Martellus map is essentially the same as on the world maps in the manuscripts of his *Insularium*, but there is a significant difference in the enclosure of these peoples. On his map at Yale, Martellus has mountains along the coast to contain Gog and Magog from the north as well (evidently he decided that the Northern Ocean was not barrier enough), but there are no such northern mountains in the world maps in his *Insularium* or on the Martellus-Rosselli map nor are there such northern mountains on Waldseemüller's map. This is additional evidence that Waldseemüller was consulting a version of Martellus's large world map different from the one at Yale, a version without those northern mountains.

While on those earlier maps Martellus only refers to the *judaei clausi*, on his map at Yale, there is also text that refers to Gog and Magog—but remarkably, that text is outside of the mountains that enclose the *judaei clausi* to the east. Martellus thus distinguishes between the *judaei clausi* and Gog and Magog. This text only approaches legibility in the multispectral images. East of the enclosing mountain range and near the northern shore of Asia are the words *provintia magogh* and to the south *ogmagog*. Further to the east, the text reads (tile C08R02):

hic sunt septem reg
na et pro maiori par
te sunt cristiani In provinti[a]

[126] For these chapters in the Naples manuscript of Marco Polo, see *Marka Pavlova z Benátek, Milion* (see note 57 in Chap. 2), Book 1, Chapter 14, pp. 18–19, and Book 1, Chapter 65, pp. 65–67, respectively; for Yule's translation, see *The Book of Ser Marco Polo* (see note 44 in Chap. 2), Book 1, Chapter 4, vol. 1, pp. 50–52, and Book 1, Chapter 59, vol. 1, pp. 284–285. Fra Mauro on his *mappamundi* has a long legend on the question of where the people enclosed by Alexander the Great were located: see Falchetta, *Fra Mauro's World Map* (see note 85 in Chap. 1), *2403, pp. 616–619.

[127] The Borgia metal *mappamundi* is described and reproduced in Destombes, *Mappemondes* (see note 10 in Front Matter), pp. 239–241 and plate 29; and the legends relevant to these peoples are transcribed by N. A. E. Nordenskiöld, "Om ett aftryck från XV:de seklet af den i metall graverade världskarta, som förvarats i kardinal Stephan Borgias museum i Velletri, Med 1 facsimile," *Ymer* 11 (1891), pp. 83–92, at p. 91. The first runs *Magog in istis duabus sunt gentes magni ut gigantes pleni omnium malorum morum. Quos judeos artaxor rex collexit de omnibus partibus persarum*; and the second: *Provincia gog in qua fuerunt judei inclusi tempore artaxorsis regis persarum*. Andrew Gow says that this map is the first in which the peoples are described as *judei clausi* in his "Gog and Magog on *Mappaemundi*" (see note 124 in Chap. 2), p. 78.

tenduth
Here there are seven kingdoms and most of the people are Christians in the province of Tangut.

There is no similar text in the same region on Waldseemüller's map. This text about Christians is certainly to be considered as separate from the toponyms referring to Gog and Magog: that is, Martellus is not asserting that Gog and Magog are predominantly Christian. But the question arises as to where Martellus got this information. The spelling *Tenduch*, which is close to Martellus's spelling *tenduth*, is used in the Naples manuscript of Marco Polo in the chapter on Tangut. And Polo says that the people in the province are mostly Christian.[128] But that chapter says nothing about seven kingdoms within the province, and I do not see how Martellus might have arrived at the conclusion that Tangut had seven kingdoms. It is possible that Waldseemüller realized Martellus's error here, and did not copy this information for that reason; it is also possible that Waldseemüller was using a different version of Martellus's map.

It is interesting that both Martellus and Waldseemüller locate Christians to the east of the mountains that surround the enclosed nations, though the textual sources of those Christians are different. The Christians on Martellus's map, just mentioned, seem to come from Marco Polo's chapter on Tangut. East of the enclosing mountains on his map, Waldseemüller has a legend that reads *hic sunt 3 ecclesie christianorum*, which comes from the preceding chapter in Marco Polo,[129] with an image of a church beside the text. On Giovanni Contarini's world map of 1506, which seems to show some influence from Martellus, there are three churches east of the mountains surrounding the enclosed races (Fig. 2.8).[130] This may perhaps be additional evidence that Waldseemüller was using a version of Martellus's large world map that was different from the one that has come down to us.

Further east on the Yale map, there is an image of a city whose name is revealed by multispectral images to be *Chatay*, that is, Cathayo, Yule's Cambaluc, i.e., modern Beijing[131]: on the world map in the London manuscript of Martellus's *Insularium*, the city of *Cathaio* is in the same location, with a legend that reads *hic dominatur cham maior tartarorum imperator*,

[128] See Marco Polo, *Marka Pavlova z Benátek, Milion* (see note 57 in Chap. 2), Book 1, Chapter 65, pp. 65–67; for Yule's translation see *The Book of Ser Marco Polo* (see note 44 in Chap. 2), Book 1, Chapter 59, vol. 1, pp. 284–285.

[129] In Marco Polo, *Marka Pavlova z Benátek, Milion* (see note 57 in Chap. 2), Book 1, Chapter 64, p. 65, the churches are basilicas: *Incole ydolatre sunt preter aliquos christianos nestorinos, qui tres ibi basilicas habent.* For Yule's translation see *The Book of Ser Marco Polo* (see note 44 in Chap. 2), Book 1, Chapter 58, vol. 1, pp. 281.

[130] The legend by the churches on Contarini's map reads *Hic sunt tres eclesie christianorum*, i.e., "Here there are three Christian churches"; the legend is transcribed and translated (differently than I have done here) in Contarini, *A Map of the World* (see note 82 in Chap. 1), p. 12.

[131] For Marco Polo's account of the city in the Naples manuscript of his work (Biblioteca Nazionale Vittorio Emanuele III, Vind. lat. 50), see *Marka Pavlova z Benátek, Milion* (see note 57 in Chap. 2), Book 2, Chapter 10, pp. 83–84; in Yule's *The Book of Ser Marco Polo* (see note 44 in Chap. 2), see Book 2, Chapter 11, vol. 1, pp. 374–379.

"Here rules the Great Khan, Emperor of the Tartars."[132] The legend below the city on the Yale Martellus map, which is only fully legible in the multispectral images of the map (tile C08R02), reads:

> *longitudo istius provintie*
> *cathay 25 dietarum et in ea*
> *dominatur cham maior tartar*
> *orum Inperator*
> *The width of this province of Cathay [is a journey of] twenty-five days, and here rules the Great Khan, Emperor of the Tartars.*

Part of this legend is very similar to that on the London *Insularium* world map, and the other part is very similar to that on the Martellus-Rosselli map: *longitudo totius provincie chati 25 dietarum.*[133] Waldseemüller copies only the first part of Martellus's legend: *longitudo tocius provincie chatij est .24. dietarum*. This legend is again consistent with Martellus's use of a manuscript closely related to Naples, Biblioteca Nazionale Vittorio Emanuele III, Vind. lat. 50, as the indication that the province of Cathay is 24 (or 25) days wide does occur in that manuscript.[134] This information does not appear in the "Z" manuscript of Pipino's Latin translation of Polo (Toledo, Biblioteca del Cabildo, 49.20, fifteenth century), for example.[135]

To the southeast on Martellus's map, there is a lake around which the remnants of buildings can barely be seen in the multispectral images of the map (C09R02); this is Quinsay, i.e., modern Hangzhou, China.[136] There are similar images of Quinsay in the same location on the world maps in the Florence and London manuscripts of Martellus's *Insularium* and on the Martellus-Rosselli map; the image on the world map in the Leiden manuscript of the *Insularium* does not include the lake. The legend

[132] Cathayo is also depicted and named on the world maps in the Florence and Leiden manuscripts of Martellus's *Insularium* and on the Martellus-Rosselli map.

[133] There is a similar legend on Giovanni Contarini's map of 1506: *longitudo provincie cathai dietarum 25*. This is further evidence that Contarini was influenced by Martellus. The legend is transcribed and translated in Contarini, *A Map of the World* (see note 82 in Chap. 1), p. 11.

[134] For the passage about the width of Cathay in the Naples manuscript of Marco Polo, see *Marka Pavlova z Benátek, Milion* (see note 57 in Chap. 2), Book 1, Chapter 63, p. 64; for Yule's translation see *The Book of Ser Marco Polo* (see note 44 in Chap. 2), Book 1, Chapter 57, vol. 1, pp. 274–276.

[135] The text of the *Z* manuscript is transcribed in Marco Polo, *The Description of the World*, trans. and ed. A. C. Moule and Paul Pelliot (London: G. Routledge, 1938), vol. 2, and with an Italian translation in Marco Polo, *Milione: redazione latina del manoscritto Z, versione italiana a fronte*, ed. Alvaro Barbieri (Parma: Ugo Guanda, 1998). For an account of the manuscript's discovery, see Homer Herriott, "The 'Lost' Toledo Manuscript of Marco Polo," *Speculum* 12 (1937), pp. 456–463; for a description of this manuscript, see Consuelo Wager Dutschke, "Francesco Pipino and the Manuscripts of Marco Polo's Travels," Ph.D. Dissertation, University of California at Los Angeles, 1993, pp. 453–456.

[136] In the Naples manuscript of Marco Polo, the description of Quinsay is in Book 2, Chapter 64: see Marco Polo, *Marka Pavlova z Benátek, Milion* (see note 57 in Chap. 2), pp. 141–142; Yule's translation is in *The Book of Ser Marco Polo* (see note 44 in Chap. 2), Book 2, Chapter 76, vol. 2, pp. 185–193, and he has extensive notes on pp. 193–200, with quotes other authors' descriptions of the city on pp. 212–215. On Quinsay also see Hallberg, *L'Extrême Orient* (see note 74 in Chap. 2), pp. 425–429, and A. C. Moule, "Marco Polo's Description of Quinsai," *T'oung Pao* 33 (1937), pp. 105–128.

near the city on the Yale map is very faint indeed, but most of it is legible in the multispectral images (C09R02):

Civitas quinsay dicitur
haberi in circuitu centum miliaria
et ignes 160 [ ] 10000
[ ]
The city Quinsay is said to be 100 miles in circumference, and to have 160 fires, 10,000 [???]...

Even setting aside the uncertainty about two words near the end of the legend, the last phrase does not make sense, for Martellus has too aggressively shortened what Marco Polo had written[137]:

In Civitate quinsay sunt ignes iuxta vulgare ytalicum seu familie tot numero, quod ad Clx tomami ascendunt, in computacione sumaria tomami uero x millia continet.

In the city of Quinsay the "fires" (according to the Italian phrase) or families total 160 tomami, *and a tomami contains a total of 10,000.*

What Marco Polo is saying is that Quinsay has a total of 1,600,000 families or households, but that does not come across in Martellus's legend. Waldseemüller has two legends about the city, and the first is related to that just cited on Martellus's map, but yet is different. The first runs:

hec civitas habet in circuitu suo 100 miliaria et in ea 12.M. pontes
This city has a circumference of 100 miles and in it there are twelve thousand bridges.

And the second:

Quinsaij civitas Mangi civitas celi in nostra ydeoma in medio lacus magnus
The city of Quinsay in Mangi; [its name means] 'City of Heaven' in our language, and in its center is a large lake.

The first of Waldseemüller's two legends about Quinsay is almost identical to that on the Martellus-Rosselli map, which reads *Quinsay hec civitas habet in circuitum .100. miliaria & in ea .12. milia pontes*, "Quinsay: this city has a circumference of 100 miles and in it there are twelve thousand bridges."

Comparing the legends about Quinsay on the Yale map, the Martellus-Rosselli map, and Waldseemüller's map provides strong evidence that Waldseemüller was consulting a different version of Martellus's large world map than the one at Yale, rather than making changes to what he found on the Yale map. Waldseemüller's legend is different from that on the Yale Martellus map, but the Martellus-Rosselli map shows that the legend that Waldseemüller used was within Martellus's repertoire. Giovanni Matteo Contarini's world map of 1506 provides additional evidence that Waldseemüller's legends come from Martellus. On Contarini's map beside the image of Quinsay, we read *Quinsai civitas celi*, "Quinsay the City of Heaven," and *Hec civitas habet in circuitum miliaria passum 100 in ea 1200*

[137]See *Marka Pavlova z Benátek, Milion* (see note 57 in Chap. 2), Book 2, Chapter 64, p. 145. Yule was working from a different text of Polo, but his translation may be compared: see *The Book of Ser Marco Polo* (see note 44 in Chap. 2), Book 2, Chapter 76, vol. 2, pp. 192.

pontes, "This city has a circumference of 100 miles and in it there are 1200 bridges."[138] The reference to the meaning of "Quinsay" does not appear on any of Martellus's other surviving maps; as Contarini shows other signs of the influence of Martellus, it seems likely that he drew both of these legends from a detailed world map by Martellus—something very similar to the Yale Martellus map. Since the evidence from Contarini indicates the existence of a Martellus map with more complete legends about Quinsay, and Waldseemüller has such legends, it seems almost certain that Waldseemüller's legends come from a different version of Martellus's large world map than the one at Yale.

Just to the southwest of Quinsay on Martellus's map, and above the banner that reads *Bangala Provi*, is a symbol for a city which is labeled *Cyanfu*, i.e., Chin-kiang-fu, now Zhenjiang, China, and below the symbol there is a very faded legend describing the city. The name and symbol for the city were evidently written in a different pigment than the legend, and thus (inconveniently) the city symbol and name, on the one hand, and the legend, on the other, do not all appear in one multispectral image. In natural light only a few of the words of the legend can be read, but more of the text is legible in multispectral images of the map (tile C09R02)[139]:

> *In ista civitate sunt due ecclesie*
> *cristianorum quas fecit marsacus [ ]*
> *istius provintie anno m ccl xxx viii*
>
> *In this city there are two Christian churches which Marsacus, [ ] of this province, built in 1288.*

This legend comes from Marco Polo's short chapter on the city of Cygianfu, and Polo indicates that Marsarchis (as the name is spelled in the Naples manuscript) was appointed praesul or protector of the city by the Great Khan.[140] Waldseemüller's legend is similar but adds the detail that Marsarchis was a Nestorian, which is not in Martellus's legend. Waldseemüller's legend reads:

> *In civitate ciansu sunt due ecclesie cristianorum quas edificauit marsarchus nestoriensis Anno domini .1288.*
>
> *In the city of Ciansu there are two Christian churches which Marsarchus the Nestorian built in the year 1288.*

This is clearly a case where either Waldseemüller was consulting a different version of Martellus's large map, one on which the legend mentioned that Marsarchis was a Nestorian, or else Waldseemüller read the

[138] The legend is transcribed and translated (differently than I have done) in Contarini, *A Map of the World* (see note 82 in Chap. 1), p. 12.

[139] The legend is clearest in Supplementary Image 2.71 (C09R02_RFTFL_cal_r90_med5_Rratio_06-12_Gratio_04-12_Bratio_01-12). There is a similar albeit shorter legend on the Martellus-Rosselli map: *In civitate cianfu sunt due eclesie cristianorum*, "In the city of Cianfu there are two Christian churches."

[140] For the passage in the Naples manuscript, see Marco Polo, *Marka Pavlova z Benátek, Milion* (see note 57 in Chap. 2), Book 2, Chapter 61, p. 139.

relevant chapter of Marco Polo and adjusted his legend to include that bit of information. The date of 1288 here may offer a clue about the manuscript of Marco Polo that Martellus was using. The Naples manuscript indicates the year as MCCLxxviii, or 1278,[141] as indeed do the F and Z manuscripts of Polo, for example,[142] and the c. 1484 edition by Leeu indicates the date as 1268.[143] The fact that the date in Martellus's legend does not agree with that in the Naples manuscript could be the result of an error by Martellus, or if a manuscript of Polo should be found that is both textually similar to the Naples manuscript and also indicates that the year Marsarchis built the churches was 1288, that manuscript would emerge as a good candidate for the one that Martellus used.[144]

South of the preceding legend on the Yale map, and northwest of the banner that reads *hec est magna provintia Mangi*, there is a legend that is invisible under natural light but appears in the multispectral images (tile C08R02).[145] Even in these images, it is very difficult to read, but the legible words are:

> *Hic sunt magna et diversa*
> *genera specierum que nunquam*
> *per[ ] ad partes*
> *[ ]*
> *Here there are many diverse kinds of spices which are never... to parts....*

It is not easy to tell which province this legend describes, as there are multiple banners indicating different provinces around it, but it probably comes from one of Marco Polo's chapters on Thebet which discusses an abundance of spices there which are not brought to Europe.[146] Waldseemüller has a very similar legend in the same location: *Hic sunt diversa genera specierum que nunquam portantur ad partes nostras*, "Here there are many different types of spices which are never brought to our regions."

---

[141] For the passage in the Naples manuscript, see Marco Polo, *Marka Pavlova z Benátek, Milion* (see note 57 in Chap. 2), Book 2, Chapter 61, p. 139.

[142] For the Latin text in the Z manuscript, see Marco Polo, *The Description of the World*, trans. and ed. A. C. Moule and Paul Pelliot (London: G. Routledge, 1938), vol. 2, p. xliv; Marco Polo, *Milione: redazione latina del manoscritto Z, versione italiana a fronte*, ed. Alvaro Barbieri (Parma: Ugo Guanda, 1998), Chapter 82, p. 198. For the passage in the F manuscript, see Marco Polo, *Il Milione, prima edizione integrale*, ed. Luigi Foscolo Benedetto (Florence: L.S. Olschki, 1928), Chapter 150, on p. 141, and for an English translation, see Marco Polo, *The Travels of Marco Polo, translated into English from the text of L.F. Benedetto by Professor Aldo Ricci* (London: G. Routledge & Sons, Ltd., 1931), p. 231.

[143] In the edition published by Leeu c. 1484, the passage is in Book 2, Chapter 61.

[144] I thank Christine Gadrat for emphasizing the importance of the date in this passage.

[145] This legend is very difficult to read; the most helpful multispectral images are Supplementary Image 2.72 (C08R02_RF+FL_cal_r90_med3_stats_from_C03R03_PC7) and Supplementary Image 2.73 (C08R02_RF+FL_cal_r90_band07+01+02+11+12+18+17_NE_Asia_light_writing_maroon_RIC5_GIC5_BPC2).

[146] See Marco Polo, *Marka Pavlova z Benátek, Milion* (see note 57 in Chap. 2), Book 2, Chapter 38, p. 115; for Yule's translation see see *The Book of Ser Marco Polo* (see note 44 in Chap. 2), Book 2, Chapter 46, vol. 2, p. 49.

South of the preceding legend, and just north of the Tropic of Cancer, there is a descriptive text between two banners: the northern one reads *hec est magna provintia Mangi* (with the word *Mangi* invisible in natural light, but visible in the multispectral images, in an extra piece of banner just below the leftmost part of the main banner), and the southern, *Ciamba provincia*. The text between these banners, which is only visible in the multispectral images, reads (tiles C08R03 and C08R02):

Hec provintia continet enim
octo regne habet linguam
propriam et canne grosse
tress spanne et longes
quindecim passus
This province contains eight kingdoms, and it has its own language, and reeds that are three hand-spans around and fifteen paces long.

Despite the presence of this text between banners referring to Mangi and Ciamba, its referent is signaled by a banner to the east that says *Thebet provincia*: Marco Polo mentions the eight kingdoms, separate language, and reeds of Thebet.[147] Waldseemüller has some of this same information on his map, but a bit to the north of where Martellus writes it, and for some reason Waldseemüller writes it twice: *hic sunt canne grosse tres spanne et longes 15 passus*, "here there are reeds three hand-spans around and fifteen paces long."

West of the preceding legend, and just north of the banner that reads *Thebet provincia* and also just north of *thebet civitas* (which is only visible in the multispectral images), is the following legend, which is only visible in the multispectral images (tile C08R02):

In omnibus istis montibus
reperiuntur lapides
preciose et in lacu
piscantur perle
In all of these mountains precious stones are found, and in the lake pearls are gathered.

This material comes again from Marco Polo, specifically his chapter on Chandu.[148] Waldseemüller has a lake in this area, *oman lacus*, in which he indicates that pearls are gathered, but as Martellus does not name the lake on his map, either Waldseemüller was using a different version of Martellus's map, or Waldseemüller did independent research about this area to depict it on his map.

North of the preceding legend, and just south of a banner that reads *churam provintia* and also just south of a city labeled *chayra* (which is

[147] Marco Polo, *Marka Pavlova z Benátek, Milion* (see note 57 in Chap. 2), Book 1, Chapter 36, p. 111, and Chapter 37, p. 113; for Yule's translation, see *The Book of Ser Marco Polo* (see note 44 in Chap. 2), Book 2, Chapter 45, vol. 2, p. 42, and Chapter 46, vol. 2, p. 49.

[148] Marco Polo, *Marka Pavlova z Benátek, Milion* (see note 57 in Chap. 2), Book 1, Chapter 38, pp. 114–115; for Yule's translation, see *The Book of Ser Marco Polo* (see note 44 in Chap. 2), Book 1, Chapter 61, vol. 1, pp. 298–304.

invisible except in the multispectral images), is the following legend, which is only legible in the multispectral images (tile C08R02):

Isti sunt ydolatre et
longitudo provincie viginti
dietarum et reperiuntur in ea
provintia calubri 10 passibus
in longitudine
These people are idolaters and the province is twenty days wide, and snakes ten paces long are found in the province.

This material comes from Marco Polo's chapter on Chairam or Carajan. The Naples manuscript does not indicate the length of the serpents in paces,[149] which is one of the few cases in which the information that Martellus supplies could not have come from that manuscript, but other versions of Polo give their length as ten paces.[150] Waldseemüller has some of this same material on his map, but a bit to the east, where a text reads *hic sunt colubri .10. p. longitudine*, "Here there are serpents ten paces long."

Northwest of the preceding legend is another one just southeast of a lake with six rivers flowing into it and one out; just south of the lake is *capaon civitas*. The city *capaon* is indicated on Waldseemüller's 1507 map, but neither the lake nor any of the material from the legend on Martellus's map appears there. So either Waldseemüller chose to omit these features and this legend for some reason, or he was consulting a different version of Martellus's large world map. Martellus's legend runs thus (part of the legend appears in tile C07R02 and the whole thing in C08R02):

isti ferunt turres
auro copertos habum
dant ita ut[?] datur misura
una auri pro .5. argenti
isti non adorant idola
sed progenitore.
The people here bear towers covered in gold, and have so much that they give one measure of gold for five of silver. They do not worship an idol, but rather their progenitors.

This text about the towers covered with gold, the exchange of gold, and the worship of ancestors is assembled from three different chapters in Marco Polo, two about the province of Mien and one about the province of Ardandam.[151]

---

[149] Marco Polo, *Marka Pavlova z Benátek, Milion* (see note 57 in Chap. 2), Book 1, Chapter 41, p. 118; for Yule's translation, see *The Book of Ser Marco Polo* (see note 44 in Chap. 2), Book 2, Chapter 49, vol. 2, pp. 76–79.

[150] See the edition of Polo in the Simon Grynaeus and Johann Huttich, eds., *Novus orbis regionum ac insularum ueteribus incognitarum* (Paris: Antoine Augerelle, 1532), Book 2, Chapt. 40, pp. 333–334; and Yule's *The Book of Ser Marco Polo* (see note 44 in Chap. 2), Book 2, Chapter 49, vol. 2, pp. 76–77 and 81.

[151] Marco Polo, *Marka Pavlova z Benátek, Milion* (see note 57 in Chap. 2), Book 1, Chapter 44, p. 126; Chapter 43, p. 125; and Chapter 41, p. 121, respectively. For Yule's translations see *The Book of Ser Marco Polo* (see note 44 in Chap. 2), Book 2, Chapter 55, vol. 2, pp. 114–115; Chapter 54, vol. 2, pp. 109–111; and Chapter 50, vol. 2, pp. 84–88.

Legends in Central Asia

On the Yale Martellus map, there are few long legends in Central Asia compared to Eastern Asia. One of these is just east of the midpoint of a hypothetical meridian joining Martellus's westward-jutting peninsula on the northern coast of Asia with Taprobana in the south. It is invisible in natural light but is legible in the multispectral images (tile C06R03):

Sacharum regio nomadum
civitates enim et villas non habet
nemora autem et speluncas
habitant
Sachara, a region of nomads, for it has no cities or towns, but they live in the woods and caves.

This comes from Ptolemy's *Geography*, Book 6, in the chapter on Sachara—and the same legend appears on the seventh map of Asia in some manuscripts and editions of Ptolemy, for example, New York Public Library MS 97, pp. 79–79; BnF MS lat. 4801, ff. 113v–114r, BnF MS lat. 4804, ff. 144v–145r,[152] and the 1482 printed edition of the *Geography*. Martellus probably took the legend from a Ptolemaic map rather than from the text, but given how frequently this legend is found on Ptolemaic maps, this information does not help us determine which Ptolemaic maps Martellus was working from. Waldseemüller has the same legend in the same spot on his 1507 map (*Sacharum regio nomadum est civitates enim non habet nemora autem et speluncas habitant*), but as this legend appears frequently on Ptolemaic maps, in and of itself, it does not provide any evidence of Waldseemüller's reliance on Martellus.

Legends in Southern Asia

There is a group of three texts legible in natural light just north of the banner that reads *India extra Gangem* and also just north of the Tropic of Cancer. The legends derive from Ptolemy, and as Waldseemüller could theoretically have found the information in them by consulting Ptolemy rather than in Martellus's map, their value for demonstrating Waldseemüller's dependence on Martellus is not high.

The westernmost of these three legends on Martellus's map reads:

Ciradia
In hac dicunt
optimum mala
bathrum fierj
Cirradia: in this [province] they say that the best cinnamon grows.

[152] Images of all of the maps in the New York Public Library manuscript of Ptolemy are available via the Digital Scriporium at http://bancroft.berkeley.edu/digitalscriptorium/, and they are also reproduced in Claudius Ptolemy, *The Geography*, trans. Edward Luther Stevenson (New York: New York Public Library, 1932; Mineola, NY: Dover, 1991). Images of all of the maps in BnF MS lat. 4801 and MS lat. 4804 are available via http://gallica.bnf.fr.

Martellus is following Ptolemy 7.2 very closely here, and Waldseemüller's legend is also very similar to Ptolemy's text: *Ciradia in hac dicunt optimum malabachinum.*[153]

The next legend to the east on Martellus's map reads *Nangalete quod paruum mundum notat*, "Nangalete, which means 'small world'." This is quite close to the text of Latin translations of Ptolemy 7.2 that were available in the late fifteenth century, but the Latin *parvum mundum*, "small world," is a mistranslation of the Greek, which is γυμνῶν κόσμος or "world of the naked."[154] Waldseemüller's legend is very similar to that of Martellus: *Nangolete quod paruum mudum notat.*

A bit further east on Martellus's map, we read *Calcitis regio in qua plura sunt metalla eris*, "The region of Calcitis in which there is much copper." Again, this is close to the text in Ptolemy 7.2, and Waldseemüller's legend is very similar: *Calcitis regio in hac multa metalla eris sunt.*[155]

Between the Tropic of Cancer and the northwestern shore of the Sinus Magnus, there is a legend which is invisible in natural light and also in several multispectral images of the area, but one processed image reveals it (tile C07R03).[156] It runs:

> *Qui hanc lestarum regionem habitant*
> *silvestres sunt et pellem habent du*
> *rum similarem hippopotamis que sagitis*
> *minime traicj potest*
> *Those who live in this region of the Lestori are wild men who have hard skin like that of a hippopotamus, which can by no means be penetrated by arrows.*

This text comes not from Marco Polo but from Ptolemy. In Book 7 of the 1482 edition of Ptolemy, in the chapter on "India extra Gangem" (sig. b7^{v}), the relevant text reads:

> *hanc lestorum regionem qui colunt silvestres esse dicuntur & in specubus habitare & pellem habere similem hippopotamis que sagittis minimi trahici potest.*
> *Those who live in this region of the Lestori are said to be wild men and live in caves and have skin like that of a hippopotamus, which can hardly be penetrated by arrows.*

There is a related legend on the 11th map of Asia in the 1482 edition of Ptolmey, but it is so incomplete as to be nonsensical: it ends with the word *similem* but does not include *hippopotamis* or any additional words. But there are manuscripts of Ptolemy that have a more complete version of this legend, for example, one of the manuscripts of the *Geography* made by

[153] For discussion of Cirradia and the cinnamon produced there, see John Watson McCrindle, *Ancient India as Described by Ptolemy* (Calcutta: Thacker, Spink & Co., 1885), pp. 219–221.

[154] For discussion of Nangalete, see McCrindle, *Ancient India* (see note 153 in Chap. 2), pp. 221 and 223.

[155] There are brief remarks on Calcitis in McCrindle, *Ancient India* (see note 153 in Chap. 2), p. 222.

[156] This text is revealed by Supplementary Image 2.74 (C07R03_RF+FL_cal_r90_med3_stats_C06R02_PCA_R2G8B22).

Martellus (Florence, Biblioteca Nazionale Centrale, Magliabechiano XIII 16, ff. 156v–157r) and also New York Public Library MS 97[157]:

> *Qui hanc regionem incolunt feri esse dicultur et in specubus habitare et pellem habere similem hippopotamis que sagittis minime traijci potest.*
>
> *Those who live in this region are said to be wild men and live in caves and have skin like that of a hippopotamus, which can hardly be penetrated by arrows.*

Thus Martellus was familiar with the legend from his work making manuscripts of the *Geography* and may have taken his legend from a manuscript Ptolemaic map, rather than directly from the text of Ptolemy.

Waldseemüller has a legend based on the same material in Ptolemy: *hanc regionem qui habitant silvestres sunt et in speluncas habitant*, "Those who live in this region are wild men and live in caves." Although this text is certainly based on the same passage in Ptolemy as the legend on Martellus's map, it is particularly clear here that Waldseemüller did not use Martellus's map at Yale as his source for this legend. Martellus does not mention the caves, but Waldseemüller does, and Waldseemüller placed his legend to the west of where Martellus did, west of the *doamas* river, while Martellus's is east of the river. In fact Waldseemüller's location for the legend is very similar to that in the 1482 edition of Ptolemy, though his wording is not particularly similar. Waldseemüller may have been using a different version of Martellus's large world map that had a different version of this legend that was located differently, but given Waldseemüller's independence of Martellus in terms of the Ptolemaic place names in the Arabian Peninsula, it is quite possible that Waldseemüller drew this legend not from Martellus, but rather from a manuscript of Ptolemy's *Geography*.

East of the northern shore of the Sinus Magnus, there is another legend that is invisible in natural light but appears in multispectral images (tile C08R03). It reads:

> *hic habitat copia*
> *magna elefantum*
> *Here live a great number of elephants.*

The source of this legend is not entirely clear; it may come from Marco Polo's chapter on the animals of Abascie, a province in southern India, where there are said to be many elephants.[158] Waldseemüller has no similar legend on his map.[159]

[157] Images of all of the maps in the New York Public Library manuscript of Ptolemy are available via the Digital Scriporium at http://bancroft.berkeley.edu/digitalscriptorium/. The maps in the manuscript are also reproduced in Claudius Ptolemy, *The Geography*, trans. Edward Luther Stevenson (New York: New York Public Library, 1932; Mineola, NY: Dover, 1991).

[158] See Marco Polo, *Marka Pavlova z Benátek, Milion* (see note 57 in Chap. 2), Book 3, Chapter 45, p. 191; for Yule's translation see *The Book of Ser Marco Polo* (see note 44 in Chap. 2), Book 3, Chapter 35, vol. 2, pp. 427–431, esp. 431 on the elephants.

[159] Waldseemüller does have a legend further to the east, in *Cyamba Provincia*, that mentions an abundance of elephants but that seems entirely separate from Martellus's legend here.

The multispectral images also reveal some legends in the southern part of the "fourth peninsula" or "tiger-leg peninsula" on the Yale Martellus map. Just south of the equator, there is a legend that is invisible in natural light but legible in a few of the multispectral images (tile C08R04).[160] It reads:

Hic non habent moenia erea
nec[?] quiquam memoratu
dignum
Here they do not have bronze walls or anything worthy of mention.

The source of this legend is Ptolemy 7.3; the text refers to the city of *Thine*, but the connection between text and city is not clear, as the city is located some distance to the southwest of the text. Waldseemüller has a similar legend in very nearly the same spot, but he manages to place the text immediately south of the city:

thine metropolis non tamen menia erea habere dicu[n]t nec quiquam memoratu dignum
The metropolis of Thina, which however they say does not have bronze walls or anything worthy of mention.

Waldseemüller's text is much closer to Ptolemy than Martellus's, so either he was consulting a different version of Martellus's map that was closer to Ptolemy, or Waldseemüller consulted Ptolemy and took the text from that book.

Further south in the peninsula, a few words of a long legend are legible in the multispectral images (still tile C08R04)[161]:

Hec [ ] est ditissi[ma]
[ ] quinque regna [ ]

These words correspond well with Waldseemüller's legend about Moabar, which he has in the same location on his map:

Hec provincia continet in se quinque regna est ditissima valde cuius incole sunt ydolatre et est maior provincia totius indie sunt nigri et maximi incantatores habent margaritas.
This province has in it five kingdoms. It is very wealthy and its inhabitants are idolaters, and it is the greatest province of all India. The people are black and great magicians, and they have pearls.

This legend comes from Marco Polo,[162] and it should be noted that in moving south from the previous legend to this one, we have moved from lands described by Ptolemy to those described by Marco Polo.

---

[160] See particularly Supplementary Image 2.75 (C08R04_RF+FL_cal_r90_med3-bands01-22_stats_from_C06R02_PC08) and Supplementary Image 2.76 (C08R04_RF+FL_cal_r90_med3-bands01-22_stats_from_C06R02_PC11).

[161] The image in which these words are the most legible is Supplementary Image 2.75 (C08R04_RF+FL_cal_r90_med3-bands01-22_stats_from_C06R02_PC08).

[162] For Polo's chapter on Moabar or Maabar in the Naples manuscript, see Marco Polo, *Marka Pavlova z Benátek, Milion* (see note 57 in Chap. 2), Book 3, Chapter 23, pp. 167–169; for Yule's translation and discussion of this chapter, see *The Book of Ser Marco Polo* (see note 44 in Chap. 2), Book 3, Chapter 16, vol. 2, pp. 331–337.

Further south on the peninsula, there are additional legible texts. The first, which is legible in natural light and also in the multispectral images (tile C08R04), runs:

hic sunt multa
animalia diu
ersa a nostris
animalibus
Here there are many animals different from our animals [i.e. European animals].

In exactly the same spot on his map, Waldseemüller has an almost identical legend: *ibi sunt multa animalia diuersa a nostris animalibus.* This information comes from Marco Polo, from text which in the Naples manuscript is in Book 3, Chapter 32, which is about the province of Comari: the text reads *Ista regio est ualde silvestris et habet animalia multum dissimilia aliis et specialiter simeas*, "This region is very wild and has animals which are very different from others, and especially apes."[163] There is a very similar passage in Book 3, Chapter 31, about Coilum (*In hac provincia animali a multa sunt cunctis aliarum regionum animantibus dissimilia*),[164] but Coilum is further to the south, and in fact Waldseemüller has a long legend in the ocean just off the southern tip of the fourth peninsula in which he quotes this sentence and several others from Polo's chapter on Coilum.

It is worth examining Waldseemüller's legend just off the southern tip of the fourth peninsula; I have added punctuation to the text for clarity[165]:

In regionibus istis penosum est vivere propter calorem nimium. in hac provincia animalia multa sunt cunctis aliarum regionum animantibus dissimilia: ibi enim sunt leones nigri totaliter, ibi sunt Papagalli albi vt nix, sunt ibi papagalli diversorum manerium. bladum nullum habent excepto riso; potum de zucharo faciunt. nudi omnes ambulant. sunt medici et astrologi; maximi luxuriosi.

In these regions it is difficult to live because of the excessive heat. In this province there are many animals which are different from all of the creatures in other regions: there are lions which are black all over, and parrots as white as snow, and parrots of many sorts. They have no grain but rice, and they make a drink from sugar [probably palm sugar]. They all go about nude. They are physicians and astrologers, and they are very wanton.

---

[163] See Marco Polo, *Marka Pavlova z Benátek, Milion* (see note 57 in Chap. 2), Book 3, Chapter 32, p. 179; the version of Polo that Yule was using was quite different here than that in the Naples manuscript, but his translation of this chapter may be consulted in *The Book of Ser Marco Polo* (see note 44 in Chap. 2), Book 3, Chapter 23, vol. 2, p. 382. Incidentally the phrasing in the Z manuscript, *in ea sunt bestie de diuersis maneriebus*, is not very similar to that in Martellus, Waldseemüller, or the Naples manuscript: see Marco Polo, *The Description of the World*, ed. Moule and Pelliot (see note 142 in Chap. 2), vol. 2, p. lxxxix. There are brief remarks on Comari in Pelliot, *Notes on Marco Polo* (see note 44 in Chap. 2), vol. 1, p. 403.

[164] For the information on Coilum, see Marco Polo, *Marka Pavlova z Benátek, Milion* (see note 57 in Chap. 2), Book 3, Chapter 31, p. 178; for Yule's translation see *The Book of Ser Marco Polo* (see note 44 in Chap. 2), Book 3, Chapter 22, vol. 2, pp. 375–376, esp. 376.

[165] For Polo's chapter on Coilum in the Naples manuscript, see Marco Polo, *Marka Pavlova z Benátek, Milion* (see note 57 in Chap. 2), Book 3, Chapter 31, pp. 178–179; for Yule's translation and discussion of this chapter, see *The Book of Ser Marco Polo* (see note 44 in Chap. 2), Book 3, Chapter 22, vol. 2, pp. 375–376.

There is no legend that corresponds to this on the Yale Martellus map; it is possible that Martellus produced a different map that included a very similar legend, but it is also possible of course that Waldseemüller composed the legend himself. Unfortunately the text does not contain good clues about which version of Marco Polo Waldseemüller might have been using: it could have been Leeu's edition of c. 1484, or it could have been a manuscript similar to the Naples manuscript, for example: Waldseemüller's phrasing is similar to that in both versions.

To the west along the coast from the preceding legend on Martellus's map is one that is only visible in the multispectral images (tile C08R04)[166]; it reads:

Ma[ ] civitas
ubi occiditur S.
thomas
apostolus
The city of Ma[ ] where the Apostle Saint Thomas was killed.

Marco Polo has a chapter on the city where St. Thomas was martyred, but he does not name the city, writing merely that it is a small city in the province of Maabar to which few merchants come, as it is not well suited to ships.[167] Martellus does not name the city on the world maps in manuscripts of his *Insularium* or on the Martellus-Rosselli map, but when he came to make his large world map, he evidently decided to identify the city with Malacca. Another candidate for the illegible city name is *Marapuria*, which is the other name for Malacca indicated by Jacob Stopel in his *Repertorium in mormam alphabeticam* (Memingen: Kune, 1519), [O7v]. Waldseemüller has the place name *Mallaqua* a short distance from the text *hic occisus est s. thomas*, which is beside a dot indicating a city. Waldseemüller does not seem completely confident in the identification of the city with the location of St. Thomas's martyrdom, but the presence of *Mallaqua* here on his map makes it very likely that that is the city that Martellus intends—a curious choice for the location of St. Thomas's martyrdom, since contrary to what Marco Polo says of that location, Malacca was frequented by merchants. Ludovico di Varthema was the first European to write about Malacca in detail, but the Portuguese had heard of the city

[166]The legend about St. Thomas is most readily legible in Supplementary Image 2.77 (C08R04_RF+FL_cal_r90_med3-bands01-22_stats_from_C06R02_PCA_R2G8B22) and Supplementary Image 2.78 (C08R04b_bands01-22_RF+FL_cal_r90_med3_stats_C06R02_PCA_R2G8B-11).

[167]The chapter on the city where St. Thomas was martyred is in Marco Polo, *Marka Pavlova z Benátek, Milion* (see note 57 in Chap. 2), Book 2, Chapter 27, pp. 173–174; for Yule's translation see *The Book of Ser Marco Polo* (see note 44 in Chap. 2), Book 3, Chapter 18, vol. 2, pp. 353–355.

when they arrived in Calicut.[168] The city's appearance on Martellus's map would seem to be the first on a European map.[169]

Directly south of the preceding legend on the fourth peninsula on Martellus's map, there is a legend which is almost all legible in natural light but more readily so in the multispectral images (tile C08R04). It reads:

Isti adorant
bovem pro deo
eorum
These men worship a cow as their god.

Waldseemüller's legend in the same spot is very similar, albeit more brief: *isti adorant boves*, "These men worship cows."[170] This comes from Marco Polo's chapter on Var or Lar.[171] Waldseemüller could have obtained this information directly from Marco Polo, but given the close correspondence in the locations of the legends and the other correspondences between the maps, it seems most likely that Waldseemüller was borrowing from Martellus here.

Legends in Northern Africa

Even in the multispectral images of the Yale Martellus map, there is not the same density of geographical information in West Africa as there is on Waldseemüller's 1507 map (Figs. 2.11 and 2.12). Given their many other congruencies, and also given the interest that Martellus evinces in the West African coast in the world maps in his *Insularium*, this is a puzzling difference between the Yale Martellus map and Waldseemüller's map. This could be an indication that Waldseemüller was using a different version of

[168] Donald F. Lach, *Asia in the Making of Europe*, vol. 1, *The Century of Discovery* (Chicago: University of Chicago Press, 1965), Book 1, p. 503; Varthema's account of Malacca is supplied in *The Travels of Ludovico di Varthema*, trans. John Winter Jones and George Percy Badger (London: Printed for the Hakluyt Society, 1863), pp. 223–228.

[169] Malacca appears on the Cantino chart of c. 1502 with a legend describing the goods available for trade there; the Cantino chart is in Modena, Biblioteca Estense Universitaria, C. G. A. 2, and is well reproduced in Nebenzahl, *Atlas of Columbus* (see note 42 in Chap. 1), pp. 35–37, and better in Armando Cortesão and Avelino Teixeira da Mota, *Portugaliae monumenta cartographica* (Lisbon: Comissão Executiva das Comemorações do Quinto Centenário da Morte do Infante D. Henrique, 1960–1962), vol. 1, plate 5, with a transcription and translation of the legend about Malacca on p. 13.

[170] The similar legend in the same location on the Martellus-Rosselli map reads *Hic abitant abrajaim adora[n]t boues*, "Here live the Brahmans, they worship cows." On the interest medieval moralists took in the story of Alexander the Great's encounter with the Bragmanni, see John B. Friedman, *The Monstrous Races in Medieval Art and Thought* (Cambridge, MA, and London: Harvard University Press, 1981; reprinted with expanded bibliography Syracuse: Syracuse University Press, 2000), pp. 169–170, who focuses on the account of the Bragmanni in Jacques de Vitry's *Historia occidentalis*; and Beverly Berg, "Dandamis: An Early Christian Portrait of Indian Asceticism," *Classica et Medievalia* 31 (1970), pp. 269–305.

[171] The chapter on Var in the Naples manuscript is supplied in Marco Polo, *Marka Pavlova z Benátek, Milion* (see note 57 in Chap. 2), Book 3, Chapter 23, pp. 169–171. For Yule's translation see *The Book of Ser Marco Polo* (see note 44 in Chap. 2), Book 3, Chapter 20, vol. 2, pp. 363–367.

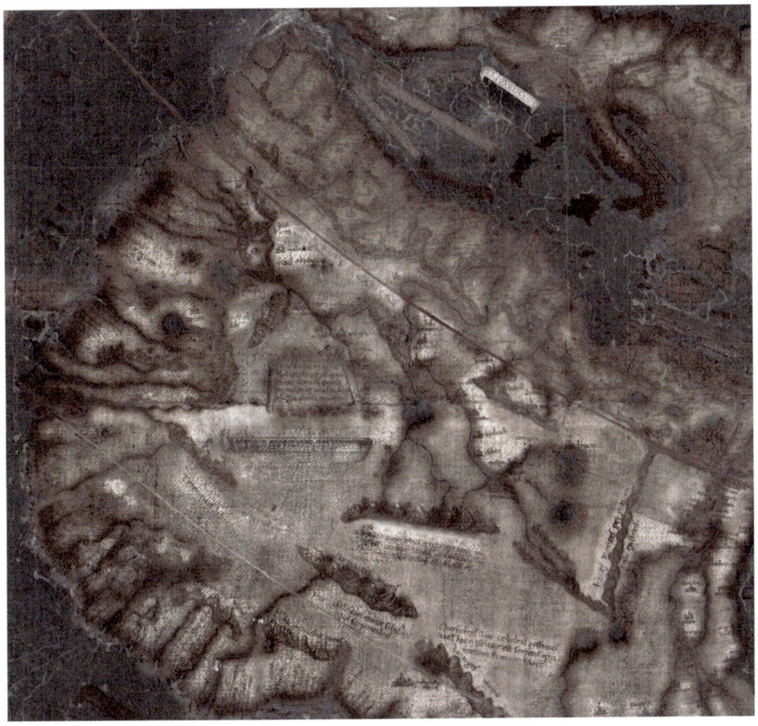

Fig. 2.11 Multispectral image of West Africa on the Yale Martellus map. (Image by Lazarus Project/MegaVision/RIT/EMEL, courtesy of the Beinecke Rare Book and Manuscript Library)

Fig. 2.12 West Africa on Waldseemüller's 1507 world map. (Courtesy of the Library of Congress)

Martellus's map that offered more information in northern Africa, or it might mean that Waldseemüller found Martellus's treatment of northern Africa to lack detail—perhaps because Martellus was more interested in southern Africa—and added a substantial amount of information to what he found on Martellus's map. There does not seem to be a way to know which was the case.

The lower number of long legends in West Africa on Martellus's map stems from a very interesting cause: he does not have any long legends from Ptolemy, whereas Waldseemüller does. Martellus does not hesitate to use long legends from Ptolemy in Western Asia and in the Indian Ocean, so it is difficult to surmise why he would not use them in West Africa. Perhaps the map is incomplete, and he did not have a chance to add the Ptolemaic legends he intended to; or perhaps he felt that in using Portuguese data for the western and southern coasts, long legends from the *Hortus Sanitatis* in the interior, and data from an *Egyptus novelo* type map for the interior of southern Africa, he was offering his viewers an updated image of the continent not available from any other cartographer. On the other hand, the contours he gives to northern and northeastern Africa still derive from Ptolemy.

In central northwestern Africa on Martellus's map, there is a banner with a legend on it that is only legible in the multispectral images. The legend reads (tile C03R03):

hic sunt magne
solitudines in quibus
sunt leones pardi
grandes et multa alia
animalia diversa a nostris
Here there are large wildernesses in which there are lions, large leopards, and many other animals different from ours.

There is a very similar legend in the same location on Waldseemüller's map that runs *Hic sunt magne solitudines et deserte in quibus sunt Leones pardi tigrides elephantes et mulcorum aliorum animalium*, "Here there are large wildernesses and deserts in which there are lions, leopards, tigers, elephants, and many other animals." That is, the two legends are very similar indeed. Elizabeth Harris has shown from typographic evidence that this legend was added to Waldseemüller's map after its initial printing.[172] Given the importance of the Martellus map to Waldseemüller even in conceiving his 1507 map, and the fact that Waldseemüller had no doubt seen this legend on one of Martellus's maps before making his 1507 map, it seems very likely that the later addition of this legend to the 1507 map was to repair damage to the woodblock, rather than to add information that had not been there previously.

[172] Elizabeth Harris, "The Waldseemüller World Map: A Typographic Appraisal," *Imago Mundi* 37 (1985), pp. 30–53, esp. 47.

To the southeast of the preceding legend, there is an unnamed range of mountains stretching west and east on Martellus's map, and just south of them, the multispectral images (tile C03R03) reveal a four-and-a-half line legend. The text must have been written with a different pigment than the preceding legend, since multispectral images that reveal the preceding legend barely show any of the present legend. However, custom processing was able to recover the text, which runs:

Sub hoc monte untia animal [ ]
est seuum stature canis [ ] longius
Inimicum canibus a quo si quis vulneratur
mures concurrent et sopra eum minguut et
sic moritur.

Beneath this mountain the panther [is found]; it is savage, with the size of a dog [but somewhat] longer, and is an enemy to dogs. If anyone is wounded by the animal mice come together and urinate on him, and thus he dies.

Similar text appears in Vincent of Beauvais's *Speculum natural*,[173] and also in the *Hortus sanitatis*[174]; given Martellus's use of the *Hortus Sanitatis* for several other legends about animals on his map, it is very likely that he was using the *Hortus Sanitatis* here rather than Vincent. And once again, Waldseemüller borrowed from Martellus. Waldseemüller's legend about the *uncia* reads:

Vntia est animal seuum stature canis parum longius a quo si quis vulneratur mures concurrent & minguut [i.e. mingunt] super eum & inmediate moritur.

The panther is a savage animal, with the stature of a dog but a little longer, and if anyone is wounded by the animal, mice come together and urinate on him and he dies immediately.

It should be noted that this cartouche on Waldseemüller's map contains a second sentence about a winged dragon that does not come from Martellus, and again either Waldseemüller was consulting a different, more elaborate version of Martellus's map, or he added to what he found there. Elizabeth Harris has shown that this legend on Waldseemüller's map, like the preceding one, was added after its initial printing.[175] The presence of source material for Waldseemüller's legend on Martellus's map renders it very likely that this legend was added to the 1507 map to repair damage to the woodblock, rather than to add information that had not been there previously.

---

[173]Vincent of Beauvais, *Speculum naturale* 19.114, "De vario & unicorne & uncia," in Vincent of Beauvais, *Speculum quadruplex; sive, Speculum maius* (Graz: Akademische Druck- u. Verlaganstalt, 1964–1965), vol. 1, cols. 1444–1445.

[174]Here is the text of the untia from the *Hortus Sanitatis*, "De animalibus," Chapter 158: *Vncia. Isidorus. Uncia est animal sevissimum: non altius cane sed longius corpore: canibus inimicum valde, predam non comedit: nisi eum in altum portet: sed quando ad arbores venit ad supremum eum ramum defert: eamque vendendo comedit. Cum autem in calore coitur existens aliquem vulneravit: mures super eum conveniunt ac super eum mingunt et moritur. Unde quidam ab ea vulneratur se per barcham in mari portari fecit: et mures ad mare veneruntur ad eum accederunt: sed propter mare non poterant. Huius animalis fel mortiferum est.*

[175]Elizabeth Harris, "The Waldseemüller World Map: A Typographic Appraisal," *Imago Mundi* 37 (1985), pp. 30–53, esp. 47.

Just north of the equator and of the *Aranges montes* (from Ptolemy 4.6) on the Yale map, there is a legend that is mostly legible in the 2010 ultraviolet image of the map but completely legible only in the multispectral images (tile C03R03). It runs:

> *Chersidras siue celidras pessimum*
> *aiat [sic] hic nasictur est [sic] serpens qui*
> *terram fumare facit*
> *They say that the chersidras or celidras—very dangerous—is found here; it is a serpent that makes the ground smoke.*

Waldseemüller has a very similar legend in exactly the same part of Africa that reads *Cherisidal sive alidras hic nascitur et est serpens qui terram fumare facit.* This serpent is discussed in a few different works, including Nicander's *Theriaca*,[176] Isidore's *Etymologiae*,[177] the anonymous thirteenth-century encyclopedia titled *Experimentator*,[178] and the *Hortus sanitatis*, "De animalibus," Chapter 36—both of these latter works merely cite Isidore. Given Martellus's use of the *Hortus Sanitatis* for information about animals elsewhere on his map, the same book was no doubt the source of his information about this serpent, and it is essentially certain that Waldseemüller took his legend from Martellus.

Southwest of the preceding legend, there is a named city with a text beneath it which is only visible in the multispectral images (tile C03R03). Because the city and its name, on the one hand, and the text on the other were written in different pigments, in the multispectral images where the city appears, the text is invisible, and vice versa. The city, which is topped with a cross, bears the name *melighon c. regalis*, and the text below the city reads:

> *hic dominatur don giovani bemu*
> *rex cenaga et bisi[ ] christianus*
> *Here rules John Bemoim, King of Senegal and [?], a Christian.*

The legend refers to Bumi Jeléen, known in Portuguese sources as João Bemoim, one of the contenders for the throne of the Kingdom of Jolof near the delta of the Senegal River, who in 1488 had traveled to Lisbon to seek support for his claims from King João II. While in Lisbon he

---

[176] See Nicander *Theriaca* 359–371 on the chersydros and 411–437 on the chelydros.

[177] See Isidore *Etymologiae* 12.4.24; in English the passage runs "The *chelydros* is a snake that is also known as the *chersydros*, as if it were *cerim* [perhaps a textual corruption], because it dwells both in the water and on land; for the Greeks call land χέρσος and water ὕδωρ. These make the earth on which they move smoke, as Macer thus describes it: 'Whether their backs froth out poison, or it smokes on the earth, where the hideous snake crawls.' And Lucan (*Civil War* 9.711): 'And the chelydri drawn along with their smoking trails.' But it always proceeds in a straight line, for if it turns when it moves, it immediately makes a sharp noise"; the translation is from Isidore of Seville, *The Etymologies of Isidore of Seville*, trans. Stephen A. Barney, W. J. Lewis, J. A. Beach, and Oliver Berghof (Cambridge, UK, and New York: Cambridge University Press, 2006), p. 256.

[178] For the text in the *Experimentator*, see Janine Deus, "Der 'Experimentator' - eine anonyme lateinische Naturenzyklopädie des frühen 13. Jahrhunderts," Dissertation, Universität Hamburg, 1998, p. 268.

converted to Christianity, and in 1489 King João sent 20 caravels under Pero Vaz de Cunha to enforce his claims, but when they arrived at the Senegal River, Cunha stabbed Bemoim to death, claiming that he was instigating treason, and returned to Lisbon.[179]

Martellus has similar legends on three of his four other surviving world maps (not on the Martellus-Rosselli map). The legend on the world map in the Florence manuscript of Martellus's *Insularium* reads *don giovanni bemu rex cenage ebazicari novus christianus 1489*, "John Bemoim, King of Senegal and Bezegiech,[180] a new Christian [in] 1489"[181]; that on the world map in the Leiden manuscript reads *don giovani bemu rex cenaga et bizicheri*, "John Bemoim, King of Senegal and Bezegiech"; and on the world map in the London manuscript, *1489 hic dominatur don giouani bemu rex cenage ebicigeri christianus novus*, "1489: Here rules John Bemoim King of Senegal and Bezegiech, a new Christian."[182]

Given that the events referred to in the text took place in 1489, the reference to them on the Yale Martellus map indicates that year as a *terminus post quem* for the map's creation (there is a later *terminus post quem*, 1491, deriving from Martellus's use of the *Hortus Sanitatis* in creating the map, as that work was first printed in that year). It is interesting that Martellus seems not to have heard the end of the story, i.e., that Bemoim was killed when he returned to Senegal: otherwise he would not have written *hic dominatur*, "here rules," on the Yale map. It is also interesting that Waldseemüller does not include this information on his 1507 map or on his 1516 *Carta marina*. It is very likely that Waldseemüller was consulting another version of Martellus's large map, but given the consistency with which Martellus includes this information on his world maps, it would be surprising if the version that Waldseemüller consulted did not include it. Thus we may have a case here in which Waldseemüller chose not to include this information on his map. Perhaps he knew that Bemoim had died and that there was thus little to be gained by recording the information; or perhaps he wanted largely to retain Ptolemaic data in northern Africa. It does not seem possible to be certain.

[179] See Rui de Pina, *Crónicas*, ed. M. Lopes de Almeida (Porto: Lello & Irmão, 1977), Chapter 37; A. Teixeira da Mota, "D. João Bemoim e a expedição portuguesa ao Senegal em 1489," *Boletim Cultural Guiné Portuguesa* 26.101 (1971), pp. 63–111; offprint published as *Série Separatas, Agrupamento de Estudos de Cartografia Antiga* 63 (1971); and António Brásio, ed., *Monumenta missionaria africana: Africa ocidental, segunda série* (Lisbon: Agência Geral do Ultramar, Divisão de Publicações e Biblioteca, 1958–), vol. 1, pp. 529–567.

[180] "Bezegiech" is the spelling of this kingdom on André Thevet's map of Africa in his *Cosmographie universelle* of 1575.

[181] My reading of the text on the Florence map differs slightly from that in Roberto Almagià, "I mappamondi di Enrico Martello e alcuni concetti geografici di Cristoforo Columbo," *La Bibliofilia* 42 (1940), pp. 288–311, at 305.

[182] My reading of the text on the London map differs slightly from that in Roberto Almagià, "I mappamondi di Enrico Martello e alcuni concetti geografici di Cristoforo Columbo," *La Bibliofilia* 42 (1940), pp. 288–311, at 305.

Just northwest of the preceding legend is another that is invisible in natural light and ultraviolet images of the map but is legible in the multispectral images (tile C03R03). It reads:

Hic basalascus nocet sibilans aliis serpentibus
Here is the basilisk which harms other serpents with its hiss.

Pliny 8.33.78 writes of the basilicus that *sibilio omnes fugat serpentes*, "it routs other serpents with its hiss," which is not quite what Martellus says, and it seems very likely that Martellus's source was the *Hortus sanitatis*, "De animalibus," Chapter 125, which says that the regulus is the same as the basilisk which is the same as the sibilius, which kills with its hiss. Waldseemüller has a legend very similar to Martellus's on his 1507 map, though a bit further to the north than Martellus: *hic basilicus nocet sibillans aliis serpent*[*ibus*], "Here the basilicus harms other serpents with its hiss." It is difficult to doubt that Waldseemüller was following Martellus here.

To the northeast there is a T-shaped mountain range west of the Nile. Beneath the right-hand bar of the "T," there is a legend which is only visible in multispectral images of the map. The beginning of the legend reads (tile C04R03):

Pelicanus avis hic
reperitur habens…
The bird pelican is found here, having…

Without being able to read more of this text about the pelican, it is impossible to be certain of the source, but it is tempting to think that Martellus was using the *Hortus sanitatis*, his favorite source for information about animals. The chapter on the pelican in that book ("De avibus," Chapter 97) does use the word *habens* in the phrase *pellem habens canam*, "having white skin," but I cannot see *pellem* or *canam* in the words that follow *habens* in the legend.[183] Waldseemüller has a closely related legend in exactly the same location: *hic pellicanus avis totus albus*, "Here is the pelican, a bird which is all white." Again it seems that either Waldseemüller was consulting a different version of Martellus's map, or else he did some research on the pelican and wrote a new legend about the animal.

To the east in the central part of the continent on Martellus's map, there is a spot where the western branch of the Nile ends in a lake, and nearby there is another lake from which the river continues to flow south. Just north of these two lakes is a short legend which is invisible in the natural light, infrared, and ultraviolet images of the map but does appear in the multispectral images (tile C03R03). It reads *hic absorbetur per tres leucas*, "Here [the river] is absorbed for three leagues." This short text is interesting, for it comes from a map called the *Egyptus novelo*, a *tabula moderna*—that is, a map based on post-Ptolemaic geographical information—of

[183]For discussion of medieval traditions about the pelican, see Victor Graham, "The Pelican as Image and Symbol," *Revue de litérature comparée* 36 (1962), pp. 233–243; and R. L. H. Lops, "Le pélican dans le Bestiaire de Philippe de Thaun," *Neophilologus* 79.3 (1995), pp. 377–387.

northern Africa that survives in three fifteenth-century manuscripts of Ptolemy's *Geography*. Martellus made heavy use of a map of this type, but of a more complete version of it than the three exemplars that have come down to us, in depicting southern Africa, as I will detail below. In the discussion of southern Africa on the Yale map below, I will provide full details about and a bibliography on the surviving *Egyptus novelo* maps,[184] but for the present I will merely cite the legends on the three surviving exemplars of the *Egyptus novelo* map that are similar to Martellus's legend about the river. On Vatican City, Biblioteca Apostolica Vaticana, MS Vat. lat. 5699, f. 125r, the legend in question reads *Absorbitur hic fluvius per tria milia*; on Vatican City, Biblioteca Apostolica Vaticana, MS Urb. lat. 277, ff. 128v–129r, it reads *Labitur occulte per 3 milia passus*; and on Paris, BnF MS lat. 4802, ff. 130v–131r, it reads *Per 3 milia sub terra labitur hic fluvius*. Thus the legend on Martellus's map is most similar to that in BAV MS Vat. lat. 5699, f. 125r. There is no similar legend on Waldseemüller's 1507 world map, but there is on his *Carta marina* of 1516, which reads *Hic absorbitur per tria miliaria*—which is very similar to that on Martellus's map.

[184] On the *Egyptus Novelo* map, see notes 5–10 in Chap. 5 below.

Chapter 3

Toponyms in Arabia, Syria, and Mesopotamia

Central Arabia Felix and Arabia Deserta, Syria, and Mesopotamia are some of the most promising areas for the investigation of place names on the Yale Martellus map, because they are some of the few parts of the map where a large number of toponyms are consistently legible, particularly in the 1959 ultraviolet images and the multispectral images. These toponyms derive from Ptolemy, and a comparison of Martellus's spellings of the toponyms in these areas with those in the editions of Ptolemy published prior to the creation of his map[1] reveals that Martellus was using a manuscript of Ptolemy, rather than a printed edition.[2] We might expect that Waldseemüller, given his use of Martellus's map as a source, would simply have copied the toponyms that he found in Arabia, Syria, and Mesopotamia on Martellus's map. But this is not the case: Waldseemüller made his own selection of Ptolemaic toponyms in these areas, so that Martellus's map has toponyms that do not appear on Waldseemüller's and Waldseemüller's map has toponyms that do not appear on Martellus's. This is a very interesting demonstration of the fact that while Waldseemüller made extensive use of Martellus as a source for descriptive texts, he generally did not depend on him for toponyms. Another instance of Waldseemüller's choice not to use Martellus as a source of place names is his copying of the vast majority of the toponyms on the western coast of Africa from Caverio's map, rather than from Martellus, which will be discussed in detail below.

These are the toponyms in the central Arabian Peninsula (central Arabia Felix) on the Yale Martellus map (tile C04R03), moving from the *montana arabie felici* south to the Tropic of Cancer, with a note whenever the toponym is different or not present on Waldseemüller's 1507 map[3]:

[1] The editions of Ptolemy's *Geography* printed prior to 1492 are those of Vincenza 1475, Bologna 1477 (which bears the mistaken date 1462), Rome 1478, Berlinghieri's verse adaptation published in Florence in 1482, Ulm 1482, Ulm 1486, and Rome 1490. For discussion of all of these editions aside from Berlinghieri's, see Wilberforce Eames, *A List of Editions of Ptolemy's Geography 1475–1730* (New York: s.n., 1886), pp. 1–7; Carlos Sanz, *La Geographia de Ptolomeo, ampliada con los primeros mapas impresos de América, desde 1507; estudio bibliográfico y crítico, con el catálogo de las ediciones aparecidas desde 1475 a 1883, comentado e ilustrado* (Madrid: Librería General V. Suárez, 1959); and Angela Codazzi, *Le edizioni quattrocentesche a cinquecentesche della "Geografia" di Tolomeo* (Milan: Goliadica, 1950). The correct date (1477) of the Bologna edition that bears the date was established by L. Sighinolfi, "I mappamondi di Taddeo Crivelli e la stampa Bolognese della Cosmografia di Tolomeo," *La Bibliofilia* 10 (1908), pp. 241–269, and see the introduction by R. A. Skelton to the facsimile edition: Ptolemy, *Cosmographia: Bologna, 1477* (Amsterdam: N. Israel, Meridian Publishing Co., 1963), p. vi.

[2] It is worth repeating here what was mentioned above in the discussion of Martellus's manuscript of Ptolemy now in Florence, namely, that he used a different manuscript of Ptolemy as a source for the Florence manuscript than he did for the Yale Martellus map.

[3] The modern names of these cities are supplied in Alfred Stückelberger et al., *Klaudios Ptolemaios Handbuch der Geographie: griechisch-deutsch* (Basel: Schwabe Verlag, 2006–2009).

C. Van Duzer, *Henricus Martellus's World Map at Yale (c. 1491)*, Historical Geography and Geosciences,
https://doi.org/10.1007/978-3-319-76840-3_3

acrna
obraca
Negram
ansara
alata
thuna
capana (not on Waldseemüller)
pharacra (not on Waldseemüller)
radi (not on Waldseemüller)
satula (sacula *on Waldseemüller)*
salma

There are a few legible toponyms in the region to the east of northern central Arabia on Martellus's map:

arra
dignna (not on Waldseemüller)
libirta (ibirta *on Waldseemüller)*

Returning now to central Arabia on Martellus's map and moving south from the Tropic of Cancer to the prominent river that cuts across Arabia:

larippa (not on Waldseemüller)
macoraba (not on Waldseemüller)
boria (perhaps an error for the city called bibanna *on Waldseemüller and in Ptolemy)*
girata (not on Waldseemüller)
castana (on Waldseemüller the toponym is further west, as cassanita regio*)*
olofia (not on Waldseemüller)
inapha
lachta (not on Waldseemüller)
chargata (not on Waldseemüller)

Now moving south of the prominent river that cuts across Arabia on Martellus's map (which does not appear on Waldseemüller's map—another difference between the maps in this region):

omanum (not on Waldseemüller)
fangala (not on Waldseemüller—and I do not find this toponym in Ptolemy either, so this is a puzzling case)
chimma (not on Waldseemüller)
sabata metropolis (not on Waldseemüller)
masomte
Rathnie
uodana (nodona *on Waldseemüller)*
madasara (not on Waldseemüller)
gorda (not on Waldseemüller)
Aragra (araregia *on Waldseemüller, and* ara regia *in various editions of Ptolemy)*
sarrium (not on Waldseemüller)

There are a couple of legible toponyms to the east of the preceding region on Martellus's map:

Iula
marimata (not on Waldseemüller)

It is also worth citing the toponyms on Waldseemüller's map in the central Arabian Peninsula that do not appear on Martellus's map, as these forcefully corroborate the independence of the sets of toponyms in this part of the two maps:

sacula
thema
Mesamnes
Maliche
carna
iolisite
Tamuce
Smirnophoros interior
mincimitosa gens
dorem
Catramonice
arabonite
Smirnoforos regio
Mafonte
Rathnie

In the 1959 ultraviolet images of the Martellus map, there are also some legible toponyms north of the *montana felicis arabie*, and in the multispectral images, there are many toponyms legible in the hinterlands of the eastern Mediterranean. As is often the case on the Martellus map, the coastal toponyms are difficult or impossible to read even in the multispectral images, but in the hinterlands in this area, there is an abundance of legible place names. I will list some of the more readily legible toponyms in the region which on Waldseemüller's map is labeled *Arabia Deserta* (it does not bear that designation on Martellus's map, another interesting difference between the two maps), with parenthetical indications about whether they appear on Waldseemüller's map—many of them do not. The fact that Waldseemüller includes the designation *Arabia Deserta* deprived him of some space that he could have devoted to the names of cities, but nonetheless, the lack of correspondence between the two maps here is striking. I transcribe from south to north (still tile C04R03):

*salima (*salma *or* salina *on Waldseemüller)*
bera (not on Waldseemüller)
thedi (not on Waldseemüller)
banatena (not on Waldseemüller)
sora
*tanba (*temna *on Waldseemüller)*
regenna (not on Waldseemüller)
cacona (not on Waldseemüller)
alamath (not on Waldseemüller)
barbarissa (not on Waldseemüller)
*gaua (*gana *on Waldseemüller)*

These comparisons of toponyms in Arabia Felix, Arabia Deserta, Syria, and Mesopotamia on the two maps abundantly confirm that as much as Waldseemüller used Martellus's map as a source for descriptive texts on his map, he was certainly not using Martellus as a source for toponyms in the Middle East.

Chapter 4

Toponyms on the Western and Southern Coasts of Africa

One of the areas on the Yale Martellus map that has elicited the greatest interest is the western and southern coasts of Africa, specifically the coastal place names. It seems likely that the Yale map, which is drawn at a larger scale than Martellus's other world maps, would contain coastal place names that his other maps do not, place names from recent voyages down the coast of Africa that would shed light on his sources. There was no expectation that Martellus's coastal African place names would have influenced Waldseemüller in making his 1507 map, for Fischer and von Wieser's statement that Waldseemüller took his coastal African place names from Nicolo de Caverio's chart of c. 1504 has been accepted without question.[1] It would certainly be interesting to know whether there is a close correspondence between the coastal African place names on the Yale Martellus map and those on Martin Behaim's globe of 1492.

In natural light, the vast majority of the coastal place names in western and southern Africa on the Yale Martellus map are not merely illegible but invisible. The 2014 multispectral images of the map render many of these coastal place names visible, at least in part. The coast itself, even in the multispectral images, is concealed in an almost impenetrable murk, and consequently the first few letters of the majority of the coastal African place names remain illegible—and several remain illegible in their entirety. Nevertheless, many of the place names are legible and enable us to draw some important conclusions.

First, the Yale Martellus map does indeed contain coastal African place names that do not appear on his other world maps, and it contains a number of names that do not appear on other contemporary maps of western and southern Africa. These names are important, previously unknown datapoints regarding early voyages of exploration down that coast. At the same time, it does not seem that the Yale Martellus map incorporates information from voyages down the coast of Africa that was not available to the cartographer when he made his other world maps. In studying the coastal African place names on all of Martellus's world maps, one reaches the conclusion that he used one detailed map of the coast of Africa in making all of his world maps and chose a different set of names from that map for each of his maps. Indeed, there are coastal African place names on his smaller world maps that do not appear on the Yale Martellus map. In

Electronic supplementary material: The online version of this chapter (https://doi.org/10.1007/978-3-319-76840-3_4) contains supplementary material, which is available to authorized users.

[1]On Waldseemüller's use of the Caverio chart as a source, see Joseph Fischer and Franz Ritter von Wieser, *Die älteste Karte mit dem Namen Amerika aus dem Jahre 1507 und die Carta marina aus dem Jahre 1516 des M. Waldseemüller (Ilacomilus)* (Innsbruck: Wagner'schen Universitäts-Buchhandlung, 1903; Amsterdam: Theatrum Orbis Terrarum, 1968), pp. 26–29. Monique Pelletier compares some of the coastal African place names on Caverio's chart and Waldseemüller's map in her article "Le Globe vert et l'oeuvre cosmographique du Gymnase vosgien," *Bulletin du Comité français de cartographie* 163 (2000), pp. 17–31.

C. Van Duzer, *Henricus Martellus's World Map at Yale (c. 1491)*, Historical Geography and Geosciences,
https://doi.org/10.1007/978-3-319-76840-3_4

particular, the place names on the Yale map do not proceed any further past the Cape of Good Hope than they do on Matellus's other world maps.

Second, a surprising result from the study of the coastal African place names on the Yale map is that a few of those names were in fact copied by Waldseemüller on his 1507 map. Even while Waldseemüller copied the vast majority of his coastal place names for Africa from the Caverio chart, he was checking Martellus against Caverio and added some place names from Martellus that do not appear in Caverio. Waldseemüller made very careful and studied use of his sources.

Third, there is a very low correspondence between the coastal place names on the Yale map and those on Martin Behaim's globe. Among all of the place names from the Yale map transcribed below, only 11 also appear on Behaim's globe (I note each of these), and there is no correspondence at all south of Monte Negro. Given that a considerable amount of other evidence indicates that Behaim used a map by Martellus as a source for his globe, this result may seem surprising, but it should not. Waldseemüller also made heavy use of a map by Martellus in making his 1507 world map—but he used a different source, the chart by Nicolo de Caverio, for most of his coastal African place names. Behaim no doubt used more than one source in making his globe, and he incorporated information about the coast of Africa from a voyage he claimed to have participated in down that same coast,[2] and this contributes to the differences between the coasts of Africa on his globe and Martellus's map, particularly in the south.

I present here the place names that I have been able to read on the western and southern coasts of Africa on the Yale Martellus map, beginning in the eastern shore of the Gulf of Guinea, just east of the island that probably represents Bioko, and proceeding southward. Some of the names in the first group are further north in Martellus's other world maps than they are on his Yale map; this is because in his other maps, he had updated the position of Africa with respect to the equator.

The brackets at the beginning of most of the entries indicate the letters that are illegible because they are lost in the "coastal murk" mentioned above. This first set of names is read from the multispectral images of tile C03R03[3]:

> [ ]picta de serr. *I see no corresponding name on Martellus's other world maps or on other contemporary maps.*

[2]See Ravenstein, "Behaim's African Voyage, 1484–1485," in his *Martin Behaim, His Life and His Globe* (see note 72 in Chap. 1), pp. 20–30. Gerald Roe Crone, "Martin Behaim, Navigator and Cosmographer: Figment of Imagination or Historical Personage?" in *Actas, Congresso Internacional de História dos Descobrimentos* (Lisbon: CIHD, 1961), vol. 2, pp. 117–132, tries to argue that the correspondence of place names on one of Martellus's world maps—presumably that in the London manuscript of his *Insularium*—and on Behaim's globe is quite close, but he seems to me to show the opposite.

[3]The multispectral that is most useful for reading these names is Supplementary Image 4.1 (C03R03_bands01-22_RF + FL_cal_r90_med3_stats_from_C06R02_PCA_R2G8B22).

[ ]a de fernando. *On the world map in the London manuscript of Martellus's* Insularium, *it is* serra di fernando, *but somewhat further north than it is on the Yale map. The Cantino map has* serra de feruando poa.[4]

[ ]de serra. *I see no corresponding name on Martellus's other world maps or on other contemporary maps.*

[ ] benedero. *The name is only visible on one multispectral image of the map.*[5] *On the Florence and London maps it is* rio di s. benedetto; *on the Leiden map it is* rio di san Benedetto; *on the Martellus-Rosselli map it is* rio de .s. Benedetto.

[ ]a. *A short name just south of the river which unfortunately is illegible.*

[ ]a nigoael [?]. *I see no corresponding name on Martellus's other world maps or on other contemporary maps.*

[ ]d meglio. *I see no corresponding name on Martellus's other world maps or on other contemporary maps.*

[ ] del principe. *On the Florence, Leiden, London, and Martellus-Rosselli maps,* angra del principe *is futher to the north—as with* serra di fernando *mentioned above, this difference of position with respect to Martellus's other maps is due to his adjustment of the position of Africa with respect to the equator on those other maps. On Behaim's globe this is* angra do principe.[6]

[ ] de barqua. *On the Florence, Leiden, London, and Martellus-Rosselli maps,* angra de barqua *is futher to the north. Also, on the Yale map this place name is adjacent to a river, but not on Martellus's other maps. On Behaim's globe it is* angra de bacco.[7]

[ ] de esares. *This name does not appear on Martellus's other world maps, but corresponds to* terra de estaeas *on Behaim's globe,* c. das esteiras *on the Cantino chart, and* cavo da esteras *on the Caverio chart.*[8]

[ ] de santa maria de nazareth. *This place name is only visible in a multispectral image that otherwise reveals few of the names on this coast*[9]*; it runs along the northern bank of the northernmost of the two rivers that reach the ocean by the headland here. The Florence map has* c. s. marie de nazaret; *the Leiden map* cavo di s. maria de nazaret, *the London map* s. maria de nazaret fluvius, *and the Martellus-Rosselli map* cavo de .s. m. de nazaret.

[ ]ues. *The name is almost invisible between two rivers, but the last three letters are legible. In all likelihood the name in full would read* c. lupi gonsalves. *On the Florence map Martellus labels this headland* c. lupi, *on the Leiden and Martellus-Rosselli maps* cavo lupi, *and on the London map* c. lupi gonsalves. *On Behaim's globe it is* Rio de lopo hizalt; *on the Cantino chart it is* c. de lope goncalves, *and on Caverio's chart it is* c. de lopo goncalvez.[10] *The cape was named for the Portuguese explorer Lopo Gonçalves, who was the first Portuguese explorer to sail south of the equator on the west coast of Africa. On the world map in the Florence manuscript of the* Insularium *and on the Martellus-Rosselli map, both of which show the equator,* c. lupi *and* cavo lupi *are a short distance south of the equator. On the Yale map the same cape is much further south of the equator because Martellus uses the unmodernized Ptolemaic position of the equator with respect to Africa, that is, with the equator cutting across the continent further north.*

[4]See Edward Luther Stevenson, *Marine World Chart of Nicolo de Canerio Januensis, 1502 (circa)* (New York: De Vinne Press, 1908), p. 99.

[5]The image that reveals *benedero* on the map Supplementary Image 2.80 (C03R03_bands01-22_RF+FL_cal_r90_difficult_cartouche_stats_PC7).

[6]Ravenstein, *Martin Behaim, His Life and His Globe* (see note 72 in Chap. 1), p. 102.

[7]Ravenstein, *Martin Behaim, His Life and His Globe* (see note 72 in Chap. 1), p. 102.

[8]Ravenstein, *Martin Behaim, His Life and His Globe* (see note 72 in Chap. 1), p. 102; Stevenson, *Marine World Chart of Nicolo de Canerio* (see note 4 in Chap. 4), p. 100.

[9]The multispectral image in question is Supplementary Image 4.1 (C03R04_RF+FL_cal_r90_med3_stats_from_C03R03_difficult_cartouche_PC7).

[10]Ravenstein, *Martin Behaim, His Life and His Globe* (see note 72 in Chap. 1), p. 102; Stevenson, *Marine World Chart* (see note 4 in Chap. 4), p. 101.

Fig. 4.1 Natural light image of the coast of Africa on the Yale Martellus map (Detail of tile C03R04). (Courtesy of the Beinecke Rare Book and Manuscript Library)

Continuing south on tile C03R04 (Figs. 4.1 and 4.2)[11]:

[ ] de uridie. *I see no corresponding name on Martellus's other world maps or on other contemporary maps.*

[ ] deserta. *I see no corresponding name on Martellus's other world maps or on other contemporary maps.*

[ ]aterina. *The Florence map has* c. de caterina, *the Leiden map* cavo de chaterina, *and the readings are similar on Martellus's other world maps. Behaim has* rio de Catherina, *the Cantino map has* C. de caterna, *and Caverio has* C. de catelina.[12] *Waldseemüller on his 1507 map mistakenly has* C de tacharma.

[ ]rito. *This name does not appear on Martellus's other world maps, but no doubt corresponds to to* sera de sto spir. *on Behaim's globe,* Serra di Sti spu *on the Cantino chart, and* Serra de S. Spirito *on the Caverio chart.*[13]

[11]The file that is most useful for reading these names is C03R04_RF+FL_cal_r90_med3_stats_from_C06R02_peninsula_bands01-22_PCA_R2G8B22.

[12]Ravenstein, *Martin Behaim, His Life and His Globe* (see note 72 in Chap. 1), p. 102; see Stevenson, *Marine World Chart* (see note 4 in Chap. 4), p. 101.

[13]Stevenson, *Marine World Chart* (see note 4 in Chap. 4), p. 101.

Fig. 4.2 Multispectral image of the coast of Africa on the Yale Martellus map (Detail of tile C03R04). (Image by Lazarus Project/MegaVision/ RIT/EMEL, courtesy of the Beinecke Rare Book and Manuscript Library)

[ ]de praia. *The word* praia *is written below the rest of the name and in a different pigment that shows as light-on-dark rather than dark-on-light in the multispectral images. This name does not appear on Martellus's other world maps, but corresponds to* Serra da praia *on the Cantino chart and* serra de praia *on the Caverio chart.*[14]

[ ] bareira. *I see no corresponding name on Martellus's other world maps or on other contemporary maps.*

[ ]a de varo martin. *The London map has* Gulfo de var martin, *and this place name corresponds to the enigmatic* G. dalvar mro *on the Cantino chart.*[15]

[ ] emsead. *The Leiden map has* cavo emscade, *but I see no corresponding name on other contemporary maps.*

[ ]a caprida. *None of Martellus's other world maps have a corresponding place name. It seems to correspond to* serra scoropoa *on Behaim's globe, the enigmatic* Serra comper *of the Cantino map, and* terra conpr… *of the Caverio map.*[16]

---

[14] Stevenson, *Marine World Chart* (see note 4 in Chap. 4), p. 101.

[15] Stevenson, *Marine World Chart* (see note 4 in Chap. 4), p. 101.

[16] Ravenstein, *Martin Behaim, His Life and His Globe* (see note 72 in Chap. 1), p. 102; Stevenson, p. 101.

[ ]a fermosa. *None of Martellus's other world maps have a corresponding name. This may correspond to* ponta formosa *on Behaim's globe and* praia fremoso *on the Cantino map.*[17]

[ ]rta bran. *I see no corresponding name on Martellus's other world maps or on other contemporary maps.*

-----. *A long place name that is illegible.*

-----. *Illegible.*

-----. *Illegible. The place name is located just north of a river and of the Tropic of Capricorn. On Martellus's other world maps this is the* rio poderoso *or a variant thereof. On Waldseemüller's 1507 map it is* rio podexo, *but this place name is not shared by Caverio, from whose map Waldseemüller took many of his place names on this coast: Caverio names the river* rio de manicogo, *and on his* Carta marina *of 1516, Waldseemüller names the river* Rio de manicongo. *So the name* rio podexo *on Waldseemüller's 1507 map represents the influence of Martellus. This is of course the Congo River, reached by Diogo Cão in 1482.*

[ ] de padrom. *The Leiden and London maps have* ponta de padron. *Caverio, who was Waldseemüller's source for many of his place names on this coast on his 1507 map, has* padrom primero, *but Waldseemüller has* ponta de padrono, *much closer to Martellus's place name. So it seems likely that Martellus was Waldseemüller's source here.*

[ ]o redondo. *The Leiden and London maps have* cavo redondo *and* cavo retondo, *respectively. Behaim has* Mue ruodo *which Ravenstein identifies with* cavo retondo *on Martellus's London map; this same place name appears on the Cantino and Caverio charts as* C. rotondo *and* cabo redondo, *respectively.*[18]

-----. *Illegible. The first river south of the Tropic of Capricorn does have a name, which is visible in a multispectral image that otherwise reveals few of the names on this coast,*[19] *but it is not quite legible.*

[ ]ra de rio. *There is no corresponding name on Martellus's other world maps. Behaim has* angra et rio de fernande, *which Ravenstein identifies with the* Rio de Fernão Vaz *on a chart attributed to Cristoforo Soligo in the Cornaro Atlas, and with the modern Dande River.*[20]

rio de fernandas. *This place name is only visible in a multispectral image that otherwise reveals few of the names on this coast*[21]*; it runs along the northern bank of the second river south of the Tropic of Capricorn. The London map also has* rio de fernandas. *This is perhaps to be identified with the* rio de fernam vas *on the Cantino map.*[22]

[ ] di maestro nicola gco da rhodi. *The word* gco *is an abbreviation for* greco, *"Greek," so "____ of Master Nicolas the Greek of Rhodes." The Florence map has* angra di mo nicola da rodi, *the Leiden map* angra di .m. nicola da rodi, *and the Martellus-Rosselli map* ang. de .m. nicola de rodi. *There is no corresponding name on the London map. The Cantino map has a* C. domestre *in this area, which seems to bear some relation to Martellus's inscription, but I do not know of any other early map—not Soligo, not Behaim, not Juan de la Cosa, not Caverio—that has a similar name in this location, except for one*[23]*: Waldseemüller on his 1507 map has* G.

---

[17] Ravenstein, *Martin Behaim, His Life and His Globe* (see note 72 in Chap. 1), p. 103; Stevenson, p. 101.

[18] Ravenstein, *Martin Behaim, His Life and His Globe* (see note 72 in Chap. 1), p. 103; Stevenson, *Marine World Chart* (see note 4 in Chap. 4), p. 102.

[19] The multispectral image in question is Supplementary Image 4.1 (C03R04_RF + FL_cal_r90_med3_stats_from_C03R03_difficult_cartouche_PC7).

[20] Ravenstein, *Martin Behaim, His Life and His Globe* (see note 4 in Chap. 4), p. 103. The Cornaro Atlas is in London, British Library, Egerton MS 73, and the chart in question is the last of three (numbers 31–33) attributed to Soligo. There is some discussion of the atlas in Armando Cortesão, *History of Portuguese Cartography* (Lisbon: Junta de Investigações do Ultramar, 1969–1971), pp. 195–201; and there is a tracing of the map with transcriptions of the place names in E. G. Ravenstein, *The Voyages of Diogo Cão and Bartholomeu Dias, 1482–88, Geographical Journal* 16.6 (1900), pp. 625–655, in the first plate.

[21] The multispectral image that reveals the name *rio de fernandas* is Supplementary Image 4.1 (C03R04_RF+FL_cal_r90_med3_stats_from_C03R03_difficult_cartouche_PC7).

[22] See Ravenstein, *The Voyages of Diogo Cão and Bartholomeu Dias* (see note 20 in Chap. 4), p. 652; and Stevenson, *Marine World Chart* (see note 4 in Chap. 4), p. 102.

[23] For lists of the place names in this area on early maps, see Ravenstein, *The Voyages of Diogo Cão and Bartholomeu Dias* (see note 20 in Chap. 4), p. 652; and Stevenson, *Marine World Chart* (see note 4 in Chap. 4), p. 102.

de magistro nicolao gco de rodi. *This is a clear case of Waldseemüller borrowing a coastal place name from Martellus in Africa, though in general Waldseemüller used Caverio as his source for the toponyms on the coasts of Africa. That said, the source of Martellus's inscription is not at all clear. Nicola da Rodi, also called Nicola il Greco and Il Palopano (or Il Palapano), was an important maker of galleys active in Venice in the first half of the fifteenth century.*[24] *Jaume Ferrar took a galley down the western coast of Africa in 1346 to search for the River of Gold,*[25] *but I am not familiar with any records of a fifteenth-century voyage down the coast of Africa in a galley, much less in one designed by Nicola da Rodi, or with him on board.*

[ ] de angra. *There is no corresponding name on any of Martellus's other world maps, or on other contemporary maps.*

[ ] capre. *The Leiden map has* insule delle capre; *Behaim has* jußula de capre, *and Cantino has* ilhas das cabras.[26] *Just south of this point on the Florence, Leiden, and Martellus-Rosselli maps there is a bay labeled* opol, *and it is surprising that it does not also appear on the Yale map, given that its larger size meant that Martellus had more room for place names. This name does not appear on the Cantino or Caverio maps, but it does appear as* opaul *on Waldseemüller's 1507 map. This is an interesting case where Waldseemüller was certainly drawing from Martellus's cartography, but not from the specific map at Yale, and thus indicates that Waldseemüller was using a different version of Martellus's large world map than that at Yale.*

[ ] de san lauren[ ]. *The Leiden map has* cavo .s. laurenzi, *and the London map has* c. s. laurenci. *Caverio has* cabo de san lorenso.[27]

[ ] serra. *There is no corresponding name on any of Martellus's other world maps, but Caverio has* cometo dacerra *at this point.*[28]

[ ]el pedroso. *Behaim has* castel poderoso de san augustin,[29] *which enables us to see that* c. s. augostini *on the London map refers to this same place. The Caverio chart has* cabo stel pedroso.

[ ]achal[ ]. *I have not found any corresponding name on Martellus's other world maps or on any early map that shows this part of the western coast of Africa, and thus the difficulty of reading the name is frustrating.*

-----. *A short name lost in the coastal murk.*

mont[ ]. *What must be the words* monte negro *appear inland in a multispectral image that otherwise reveals few of the names on this coast*[30]*; it bears remarking that Martellus places this place name inland on his other world maps as well. The illustration of Monte Negro is very similar to those on Martellus's other maps. South of this point there is no correspondence between the place names on the Yale Martellus map and those on Behaim's globe.*

[ ] alta. *The Florence, Leiden, London, and Martellus-Rosselli maps have* terra alta*; this name does not appear in this area on other early maps.*

---

[24]On Nicola da Rodi, see Franco Rossi, "L'arsenale: I quadri direttivi"; in Gino Benzoni and Antonio Menniti Ippolito, eds., *Storia di Venezia: dalle origini alla caduta della Serenissima* (Rome: Istituto della Enciclopedia italiana, 1991–), vol. 5, *Il rinascimento: società ed economia*, eds. Alberto Tenenti and Ugo Tucci, pp. 593–639, esp. 604–609; and Mauro Bondioli and Gilberto Penzo, "Teodoro Baxon e Nicola Palopano proti delle galee sottili. L'Influsso greco nelle construzioni navali veneziane della prima metà del XV secolo," *Archeologia delle Acque* 1 (1999), pp. 67–80.

[25]On Ferrer's voyage, see Richard Hennig, *Terrae Incognitae* (Leiden: E. J. Brill, 1944–1956), vol. 3, pp. 284–289; and G. Llompart i Moragues, "L'identitat de Jaume Ferrer el Navegant (1346)," *Memòries de la Real Acadèmia d'Estudis Històrics, Geneològics i Heràldics de Balears* 10 (2000), pp. 7–20. There is a legend about Ferrer's voyage and also an image of him in the Catalan Atlas of 1375; the atlas is in Paris, Bibliothèque nationale de France, MS Espagnol 30; and the legend about Jaume Ferrer's voyage is transcribed and translated in the commentary volume that accompanies a facsimile of the atlas, *Mapamundi del año 1375* (Barcelona: S.A. Ebrisa, 1983), p. 33.

[26]See Ravenstein, *Martin Behaim, His Life and His Globe* (see note 72 in Chap. 1), p. 102; and Stevenson, *Marine World Chart* (see note 4 in Chap. 4), p. 101.

[27]Stevenson, *Marine World Chart* (see note 4 in Chap. 4), p. 102.

[28]Stevenson, *Marine World Chart* (see note 4 in Chap. 4), p. 102.

[29]Ravenstein, *Martin Behaim, His Life and His Globe* (see note 72 in Chap. 1), p. 103.

[30]The multispectral image in question is Supplementary Image 4.1 (C03R04_RF+FL_cal_r90_med3_stats_from_C03R03_difficult_cartouche_PC7).

Fig. 4.3 Natural light image of the coast of Africa on the Yale Martellus map (From tile C03R05). (Courtesy of the Beinecke Rare Book and Manuscript Library)

[ ]llo pranbero. *This reading is not certain. Caverio and Waldseemüller have* Caleta primera *here,*[31] *but the name on Martellus's map does not look like that to me. If this reading is close to correct, there is no corresponding name on Martellus's other world maps or on other early maps of this area.*

gulfo de aldeas. *The Florence map has* g. de aldeas; *the Leiden map has* golfo de aldeas; *Cantino has* G. das aldeas, *and Caverio* Corffo daalgas.[32]

Continuing south on tile C03R05 (Figs. 4.3 and 4.4):

isla et punta de .s. domengo. *The Leiden map has* ponta di .s. domenicho; *this name does not appear in this area on other early maps. St. Dominic's feast day was August 4, so the island and point here may have been discovered on that day. The following stretch of coastline on the Yale Martellus map has an elevated density of place names.*

[ ]sar[ ]. *Unfortunately the name is mostly illegible.*

darea. *This seems to be a mistaken repetition of part of the next place name.*

p[ ]a darea. *Thre is no corresponding place name on any of Martellus's other world maps; Cantino has* punta darca *here, and Caverio* plaia darea.[33]

c[ ] de manga. *There is no corresponding name on Martellus's other world maps or on other early maps of this area. The section of the coast from here to the east has an unusually high number of place names that do not occur on other early maps.*

[ ]agia [ ]erde. *The Florence, Leiden, and Martellus-Rosselli maps have* piagia verde, *and the London map has* p. verde; *Caverio has* plaia verde.

[ ]ao preto. *Juan de la Cosa has* Cabo preto.[34] *This spelling of the name does not appear on any other of Martellus's maps, or on other early maps of this area.*

[ ] de naisas. *There is no corresponding name on Martellus's other world maps or on other early maps of this area.*

[ ] de ramos. *There is no corresponding name on Martellus's other world maps or on other early maps of this area.*

[ ] de gad. *There is no corresponding name on Martellus's other world maps or on other early maps of this area.*

[31] Stevenson, *Marine World Chart* (see note 4 in Chap. 4), p. 103.

[32] Stevenson, *Marine World Chart* (see note 4 in Chap. 4), p. 103.

[33] Stevenson, *Marine World Chart* (see note 4 in Chap. 4), p. 103.

[34] Hugo O'Donnell y Duque de Estrada, *El mapamundi denominado 'Carta de Juan de la Cosa'* (Madrid: Egeria, 1992), p. 159; also see Ravenstein, *The Voyages of Diogo Cão and Bartholomeu Dias* (see note 20 in Chap. 4), p. 653.

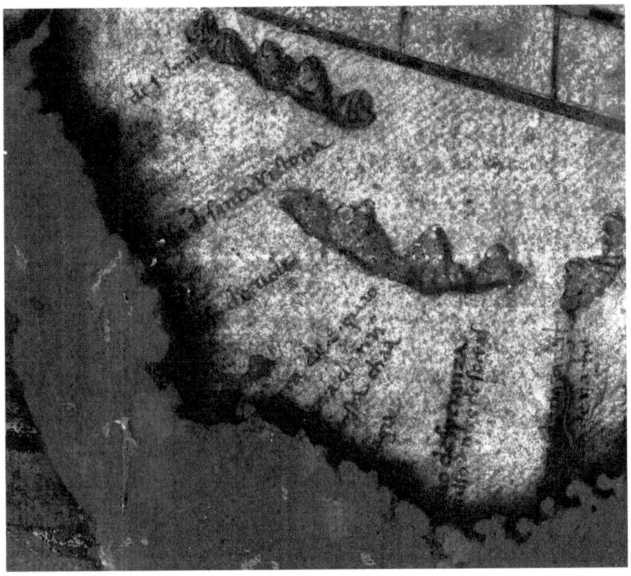

Fig. 4.4 Multispectral image of the coast of Africa on the Yale Martellus map (Detail from tile C03R05). (Image by Lazarus Project/MegaVision/RIT/EMEL, courtesy of the Beinecke Rare Book and Manuscript Library)

[ ] de balena. *The Florence, Leiden, and London maps have* golfo de balena, *and the Martellus-Rosselli has* gulfu de balena. *Juan de la Cosa has* C de la ballena,[35] *but this name does not appear on the Cantino or Caverio maps. Like the place name* ao preto *above, this name indicates some connection between Juan de la Cosa's sources and Martellus's.*

[ ]o lertam. *There is no corresponding name on Martellus's other world maps or on other early maps of this area—except on Juan de la Cosa's map, who has* C do leitom *just south of* C de la ballena.[36] *Like the preceding place name, this one indicates a connection between Juan de la Cosa's sources and those of Martellus.*

[ ] de damata. *There is no corresponding name on Martellus's other world maps or on other early maps of this area.*

[ ]a brava. *There is no corresponding name on Martellus's other world maps or on other early maps of this area. The London map does have the place name* arena brava, *but well up the coast, just south of Monte Negro.*

[ ] de p[ ]dron. *The London map has* c. de padron *here; the Cantino chart has* C. do padro, *and Caverio has* .c. de padrom.[37] *There is a very significant difference in the order of the place names on Martellus's world maps in this area. The order is fairly consistent among the Florence, Leiden, and Martellus-Rosselli maps, but different on the London map. Specifically, the London map has* golfo de balena, *then* c. de padron *and* piagia de sardingha, *followed by* serra parda, c. de s. spirito, seramal, *and* cavo de arenas. *On the other maps, there are five place names between the* golfo de balena *and* c. de padron *(rather than zero), and those place names include three—*serra parda, c. de s. spirito, *and* cavo de arenas*—that come after* c. de padron *on the London map. The problem seems to be in the London map, as both* golfo de balena *and* cavo de arenas *are repeated further down the coast, and there is a repetition of* serra a paradam *as well.*

-----. *An illegible place name.*

[35] Ravenstein, *The Voyages of Diogo Cão and Bartholomeu Dias* (see note 20 in Chap. 4), p. 653, who incorrectly lists the name as *Golfo da baleia*. The correct reading is supplied by Hugo O'Donnell y Duque de Estrada, *El mapamundi denominado 'Carta de Juan de la Cosa'* (Madrid: Egeria, 1992), p. 159.

[36] The place name *C do leitom* on Juan de la Cosa's map is not listed by Ravenstein, *The Voyages of Diogo Cão and Bartholomeu Dias* (see note 20 in Chap. 4), p. 653, but it is supplied by O'Donnell y Duque de Estrada, *El mapamundi denominado 'Carta de Juan de la Cosa'* (see note 34 in Chap. 4), p. 159.

[37] Stevenson, *Marine World Chart* (see note 4 in Chap. 4), p. 104.

[ ] de sard[ ]. *The Florence and Martellus-Rosselli maps have* praia de sardigna *here, and the London map has* piagia de sardingha. *I do not see a similar name on other early maps of this area.*

-----. *An illegible place name along the west bank of a river. I suspect that it was* serra parda, *as this name is adjacent to what seems to be the corresponding river on the Leiden, Martellus-Rosselli, and London maps.*

[ ] de [ ]ate. *The reading* de palmate *is tempting. The few letters that are legible here offer little to go on, but I do not see any corresponding name on Martellus's other world maps or on other early maps of this area.*

Golfo di santa Victoria. *This name does not appear on any of Martellus's other world maps. Juan de la Cosa has* G de s vitoria, *and the Cantino map has* G. de Sam vitoria, *while the Caverio map has* Gorffo de .S. antoni *in the same location.*[38]

[ ]olfo de volta. *The Leiden map has* golfo de volta, *and the London map has* cavo de volta. *There is no corresponding name on the Juan de la Cosa, Cantino, or Caverio maps, or on Waldseemüller's 1507 map.*

[ ]erra de s. grazo. *There is no corresponding name on Martellus's other world maps or on other early maps of this area.*

[s]erra de reis. *The word* reis *is very difficult to make out, but this reading is confirmed by the Leiden map, which has* serra de reis *here. Juan de la Cosa has* sierra dos Rey, *and Caverio has* os reis.[39] *This name had been given by Bartholomeu Dias.*

[ ]sta chaa. *The Florence, Leiden, and Martellus-Rosselli maps have* acostaihaa, *and the London map has* acostaiha. *There is no corresponding name on other early maps of this area.*

[ ]agia. *There is no corresponding name on Martellus's other world maps or on other early maps of this area.*

Cavo desperanza. *The Cape of Good Hope, discovered and named by Bartholomeu Dias in 1488.*[40]

Gulfo dentro de serras. *The Leiden map has* gulfo de sera, *and the London map has* gulfo dento delle serre. *There is no corresponding name on Martellus's other world maps or on other early maps of this area. This name was given by Bartholomeu Dias. This is interesting evidence that Martellus had access to information about Dias's voyage that was not available to other cartographers.*

serra de nazarith. *There is no corresponding name on Martellus's other world maps or on other early maps of this area.*

[ ]o de vache. *The Leiden and London maps have* rio delle vache, *while the Martellus-Rosselli map has* rio delle nache. *No other early map of this area has a similar place name. The name probably derives from Bartholomeu Dias's* Bahia dos vaqueiros, *"River of the Cowherds."*

Continuing to the east on tile C04R05:

serra de s thomaso. *The Leiden map has* serra de .s. tamasio. *There is no corresponding name on other early maps of this area. The feast day of St. Thomas the Apostle was on December 21, so this name may represent a discovery made on that day, but of course there are other saints named Thomas, and other reasons the name may have been given.*

callado. *The Leiden map has* callado, *and the London map has* cavo dalhado. *The Cantino chart has* C. talhado *and the Caverio chart has* cabo talcado.[41]

[38] O'Donnell y Duque de Estrada, *El mapamundi denominado 'Carta de Juan de la Cosa'* (see note 34 in Chap. 4), p. 159; Stevenson, *Marine World Chart* (see note 4 in Chap. 4), p. 105.

[39] O'Donnell y Duque de Estrada, *El mapamundi denominado 'Carta de Juan de la Cosa'* (see note 34 in Chap. 4), p. 159; Stevenson, *Marine World Chart* (see note 4 in Chap. 4), p. 105.

[40] The naming of the Cape of Good Hope has sometimes been falsely attributed to King João II of Portugal; for evidence that it was named by Bartolomeu Dias, see Eric Axelson, "The Dias Voyage, 1487–1488: Toponymy and Padrões", *Revista da Universidade de Coimbra* 34 (1988), pp. 29–55, at 45; the article was also published as an offprint by the Centro de Estudos de História e Cartografia Antiga, Serie Separatas, no. 189 (1988).

[41] Stevenson, *Marine World Chart* (see note 4 in Chap. 4), p. 106.

[ ]de exo[ ]da. *Unidentifiable. The names from this point east have suffered some damage, perhaps scuffing, and they are very difficult to read, even in the multispectral images.*[42]

[ ]d[ ]a[ ]. *Unidentifiable. It is tempting to try to identify this name with the* golfo de pastori *that consistently appears on Martellus's other world maps, but I cannot convince myself of that identification based on what is visible in the multispectral images.*

[ ]fo d[ ]ocha. *Certainly for* golfo de rocha, *corresponding to* golfo de roca *on the Leiden map.*

[ ]giaan. *Given the presence of the "gi" near the end of the name, it is tempting to try to identify it with the* padram de s. giorga *and similar names here on Martellus's other maps, but the letters at the end of the word do not support this identification.*

[ ]alagos. *Unidentifiable. There is no similar name in this area in any of Martellus's other world maps.*

[ ]ha de f[ ]. *For* ilha de fonte, *which is the easternmost place name on this coast on all of Martellus's other world maps, and this same point is proclaimed the easternmost point of the Portuguese voyages in a text (discussed earlier) in a cartouche just to the south,* huc usque columpnam et ilha de fontj sive penedo pervenit classis Johan[nis] secundi, *"To this point, to the column and the Isle (or Rock) of the Fountain came the fleet of John the Second."*

The Yale Martellus map thus turns out to be an important source on the early Portuguese exploration of the western coast of Africa, for it contains numerous place names that do not occur on other surviving maps. It is tempting to think that some of these names derive from maps from the 1487–1488 voyage of Bartolomeu Dias, but precisely because the names do not occur in other sources, it is not possible to be certain of their source. The fact that the map does not include post-Dias data on the coast of Africa is also helpful in terms of dating the map: given Martellus's pride in using the latest information on the Portuguese discoveries, it must have been made before information about Vasco da Gama's voyage to India (1497–1498) was diffused.

[42]These names are closest to being legible in Supplementary Image 4.3 (C04R05_bands01-22_RF+FL_cal_r90_stats_from_peninsula+cartouche_bands01-22_PC08), Supplementary Image 4.4 (C04R05_bands01-22_RF+FL_cal_r90_stats_from_peninsula+cartouche_bands01-22_PC11), and Supplementary Image 4.5 (C04R05_RF+FL_cal_r90_bands_05+06+12+16_lake_name_cyan_PCA_stats_from_C05R05_PC2).

Chapter 5

Southern Africa and the *Egyptus Novelo* Maps

Southern Africa turns out to be one of the most remarkable and surprising regions on the Yale Martellus map, but none of its details can be seen in natural light: it can only be studied though images taken with light beyond the visible spectra. The ultraviolet images of the map made in 1959 and 2010 reveal many of the details in the interior, and the 2010 infrared photographs show that Martellus wrote the names of the rivers in southern Africa with the same pigment he used for the water of the rivers, rather than the paint he usually used for toponyms and legends. But the 2014 multispectral images are by far the best tools for studying this part of the map: they clearly reveal named cities, mountain chains, and the full extent of the southern part of the Nile river system, which reaches to the southeastern part of the continent (Fig. 5.1).

On the Yale Martellus map southern Africa is shaped like a foot or shoe, as Skelton suggested in his study of the map, with the heel to the southwest, and the toe consisting of the peninsula that juts to the east.[1] As noted above, Martellus experimented with the shape of southeastern Africa in his different world maps, showing an eastward-jutting peninsula on his map now in Florence (see Fig. 1.2), but not on his map now in Leiden (see Fig. 1.3) or the printed map he produced with Rosselli (see Fig. 1.6), and yet showing such a peninsula again on his map now in London, though obscuring its full extent (see Fig. 1.4), and showing its full extent on his Yale map. All of Martellus's surviving maps seem to have been made before Vasco da Gama circumnavigated Africa in 1497–1498, so Martellus's depiction is at least to some extent speculative. But what are the sources of this peninsula?

Skelton suggests that the "toe" of the southern African shoe is a remnant of the land bridge that Ptolemy believed joined southern Africa to Southern Asia, and this seems likely, particularly as the shape of the fourth peninsula in Asia on Martellus's maps, which juts far to the southwest toward the southern tip of Africa, also seems to be a remnant of the Ptolemaic land bridge.[2] It is tempting to see Martellus's indecision about the peninsula jutting eastward from southern Africa as a reflection of his wondering about how far data from Ptolemy could be trusted at a time when explorers were finding a number of errors not only in Ptolemy's *Geography* but also in other classical geographical authorities.[3] This entire unbroken land bridge

---

Electronic supplementary material: The online version of this chapter (https://doi.org/10.1007/978-3-319-76840-3_5) contains supplementary material, which is available to authorized users.

[1] See R. A. Skelton, "World Map by Henricus Martellus Germanus, c. 1489, at Berne," January 10–17, 1960 (see note 2 in Front Matter), p. 6.

[2] On the Ptolemaic land bridge between southern Africa and Asia, see note 89 in Chap. 1 above.

[3] On complaints of Renaissance scholars about errors in classical geographical texts, see Reijer Hooykaas, *Humanism and the Voyages of Discovery in Sixteenth Century Portuguese Science and Letters* (Amsterdam and New York: Noord-Hollandsche U.M., 1979); and Thomas Goldstein, "Geography in Fifteenth-Century Florence," in John Parker, ed., *Merchants and Scholars: Essays in the History of Exploration and Trade* (Minneapolis: University of Minnesota Press, 1965), pp. 9–32; reprinted in Felipe Fernández-Armesto, ed., *The European Opportunity* (Aldershot, Great Britain, and Brookfield, VT: Variorum, 1995), pp. 1–22.

C. Van Duzer, *Henricus Martellus's World Map at Yale (c. 1491)*, Historical Geography and Geosciences, https://doi.org/10.1007/978-3-319-76840-3_5

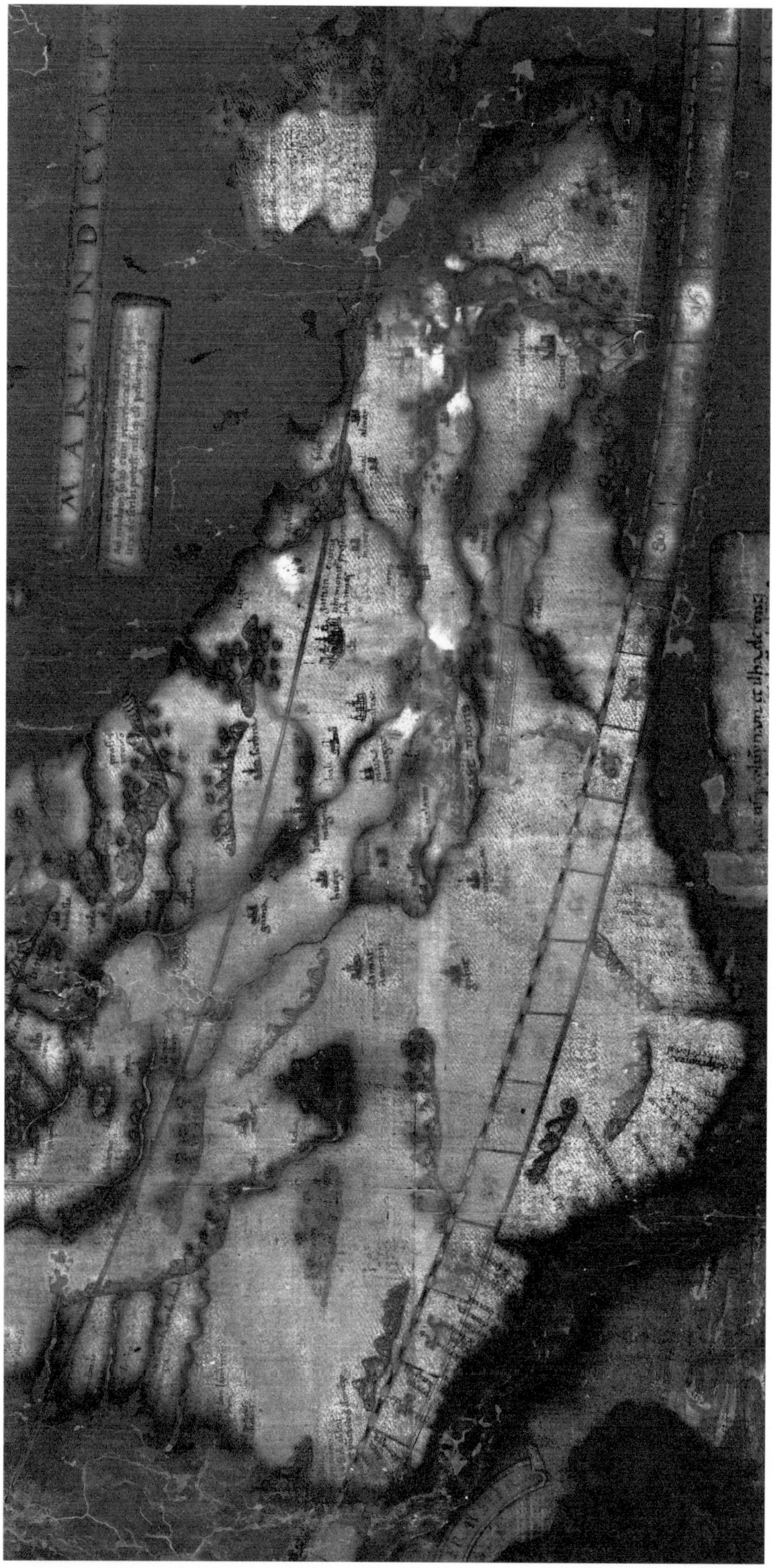

Fig. 5.1 *Multispectral image of southern Africa on the Yale Martellus map. (Image by Lazarus Project/MegaVision/RIT/EMEL, courtesy of the Beinecke Rare Book and Manuscript Library)*

is depicted on one fourteenth- and two fifteenth-century *mappaemundi*,[4] and apparent remnants of the land bridge similar to what we see in Africa on the Yale Martellus map—that is, a peninsula jutting eastward from southeastern Africa—appear on many fifteenth-century maps.

Let us look more closely at the interior details of southern Africa on the Yale Martellus map in a multispectral image (see Fig. 5.1), which reveals much that is obscure in natural light images. A striking difference between Martellus map at Yale and his earlier world maps is immediately evident: while the interior of southern Africa is essentially empty on Martellus's maps now in Florence, Leiden, and London, and his map printed by Rosselli (see Figs. 1.2, 1.3, 1.4, and 1.6), southern Africa is full of detail on the map at Yale. On Martellus's other world maps, the Nile river system is traditionally Ptolemaic: the river rises from the Mountains of the Moon about two-thirds of the way down the continent, flowing from these mountains into two lakes, and then northward toward the Mediterranean. On the Yale Martellus map, the northern part of the Nile is traditional from a large mountain range about two-thirds of the way down the continent to the Mediterranean, but the river system continues south of the mountains, stretching with many branches all the way to the southeastern corner of the continent. Moreover, the Yale Martellus map has many mountain ranges and named cities in southern Africa. Upon close examination, it seems that the eastward-jutting peninsula is cut off from mainland Africa by a narrow channel with mountains on either side of it, and near the eastern tip of the land, there is a lake. It is very clear that in creating his Yale map, Martellus used a source or sources about southern Africa which he did not use in making his other earlier maps; it is reasonable to take this as confirmation that his Yale map is later than his other world maps.

I will show that the map that Martellus used as a source for his new depiction of southern Africa on his Yale map was a variant of the *Egyptus Novelo* map. The *Egyptus Novelo* map survives in three manuscripts of

[4]The fourteenth-century *mappamundi* that shows the land bridge is in Paris, Bibliothèque nationale de France, MS lat. 6556, f. 7v, c. 1350, on which see Destombes, *Mappemondes* (see note 10 in Front Matter), pp. 176–177, no. 50.17, illustrated in Youssouf Kamal, *Monumenta cartographica Africae et Aegypti* (Cairo, 1926–51), vol. 3, fasc. 3, f. 820; and *The World Encompassed: An Exhibition of the History of Maps Held at the Baltimore Museum of Art October 7 to November 23, 1952* (Baltimore: Trustees of the Walters Art Gallery, 1952), no. 35, plate 20, Fig. 1. The two fifteenth-century *mappaemundi* that show the Indian Ocean surrounded by land are (1) Venice, Biblioteca Nazionale Marciana, Cod. It. VI, 24 (6111), f. 67r, which was in an anonymous fifteenth-century geographical tract but is now missing, on which see Destombes, *Mappemondes* (see note 10 in Front Matter), p. 188, no. 51.36, and p. 319, and for a hand-drawn reproduction, Konrad Kretschmer, *Die Entdeckung Amerika's in ihrer Bedeutung für die Geschichte des Weltbildes* (Berlin: W. H. Kühl, 1892), Atlas, plate 3, no. 13, whence it is reproduced in Giuseppe Caraci, "Cartografia," in *Enciclopedia Italiana* (Rome: Istituto Giovanni Treccani, 1929–1961), vol. 9, plate 71, no. 13, and there is text in English about the map in the reprint of Kretschmer's atlas, Konrad Kretschmer, *Die historischen Karten zur Entdeckung Amerikas: Atlas nach Konrad Kretschmer*, ed. Oswald Dreyer-Eimbcke (Frankfurt am Main: Umschau, 1991), p. 62; and (2) an unsigned and undated copper-plate Venetian world map of ca. 1485 in Vienna, Österreichische Akademie der Wissenschaften, Collection Erich Woldan, reproduced and discussed in Erich Woldan, "A Circular, Copper-Engraved, Medieval World Map," *Imago Mundi* 11 (1954), pp. 12–16.

Ptolemy's *Geography* that were painted between 1469 and about 1480 by Pietro del Massaio in Florence, and they combine a Ptolemaic depiction of Egypt and the northern Nile system with a map of Ethiopia (i.e., Africa south of Egypt) that was based on a new, post-Ptolemaic description of the regions,[5] perhaps communicated by Ethiopian delegates to the Council of Florence in 1441.[6] The three manuscripts of Ptolemy's *Geography* that include this *tabula moderna* or modern map are[7]:

1. Vatican City, Biblioteca Apostolica Vaticana, MS Vat. lat. 5699, f. 125, dated 1469, where the map is titled *Aegyptus* cum *Ethiopia moderna*[8]
2. Vatican City, Biblioteca Apostolica Vaticana, MS Urb. lat. 277, ff. 128v-129r, dated 1472, where the map is titled *Descriptio Egypti Nova*[9]
3. Paris, Bibliothèque nationale de France, MS lat. 4802, ff. 130v-131r, made c. 1475–80,[10] where it is titled *Egyptus Novelo* (see Fig. 5.2)

The best studies of the *Egyptus Novelo* maps are those by Bertrand Hirsch, in an article and his Ph.D. dissertation.[11] In these works Hirsch

[5]For brief general discussions of the *Egyptus novelo* map, see Laura Mannoni, *Una carta italiana del bacino del Nilo e dell'Etiopia del secolo XV* (Rome: s.n., 1932) = Pubblicazioni dell'Istituto di geografia della R. Università di Roma, Ser. B, num. 1; Bertrand Hirsch, "Cartographie et itinéraires: Figures occidentales du nord de l'Ethiopie aux XVe et XVIe siècles," *Abbay* 13 (1986) pp. 91–122; and Gianfranco Fiaccadori, "Egyptus Novelo," in Siegbert Uhlig, ed., *Encyclopaedia Aethiopica* (Wiesbaden: Harrassowitz, 2003), vol. 2, pp. 246–247. Also see O. G. S. Crawford, "Some Medieval Theories about the Nile," *Geographical Journal* 114 (1949), pp. 6–23, esp. pp. 8–19 and W. G. L. Randles, "South-East Africa as Shown on Selected Printed Maps of the Sixteenth Century," *Imago Mundi* 13 (1956), pp. 69–88, esp. 73–76 and the reproduction between pp. 69 and 70.

[6]So Mannoni, *Una carta italiana* (see note 5 in Chap. 5), and Anthony O'Mahony, "Between Islam and Christendom: The Ethiopian Community in Jerusalem Before 1517," *Medieval Encounters* 2.2 (1996), pp. 140–154; also see Biondo Flavio, *Historiarum ab Inclinatone Romanorum, Quartae Decadis Liber Secundus*, in B. Nogara, ed., *Scritti inediti e rari di Biondo Flavio* (Rome: Tipografia poliglotta Vaticana, 1927), pp. 3–28, esp. sects. 35–36, pp. 21–22.

[7]All three manuscripts are described by Germaine Aujac, "Le peintre florentin Piero del Massaio et la 'Cosmographia' de Ptolémée," *Geographia antiqua* 3–4 (1994–1995), pp. 187–204, and Louis Duval-Arnould, "Les manuscrits de la *Géographie* de Ptolémée issus de l'atelier de Piero del Massaio (Florence, 1469–vers 1478)," in Didier Marcotte, ed., *Humanisme et culture géographique à l'époque du Concile de Constance autour de Guillaume Fillastre: Actes du Colloque de l'Université de Reims, 18–19 novembre 1999* (Turnhout: Brepols, 2002), pp. 227–244.

[8]There is a brief description of BAV MS Urb. lat. 5699 in Adriana Marucchi, "Codici di Niccolò Perotti nella Biblioteca Vaticana," *Humanistica Lovaniensia* 34A (1985), pp. 99–125, at 118–119.

[9]BAV MS Urb. lat. 277 has been published in facsimile as *Codex Urb. Lat. 277: La cosmografía de Claudius Ptolomeus: transcrito entre los años 1472 y 1473* (Madrid: Encuentro, 1983), with a commentary volume by Arthur Dürst. The *Descriptio Egypti Nova* map in this manuscript is briefly described by Dürst in p. 68 of his commentary.

[10]In earlier literature a date of 1456 was often assigned to BnF MS lat. 4802, but this is corrected by Bertrand Hirsch, "Cartographie et itinéraires: Figures occidentales du nord de l'Ethiopie aux XVe et XVIe siècles," *Abbay* 13 (1986) pp. 91–122, at 98; Aujac, "Le peintre florentin Piero del Massaio" (see note 7 in Chap. 5), and Duval-Arnould, "Les manuscrits de la *Géographie* de Ptolémée" (see note 7 in Chap. 5). The maps in BnF MS lat. 4802 have been reproduced in facsimile in *Géographie de Ptolémée: traduction latine de Jacopo d'Angiolo de Florence; reproduction réduite des cartes et plans du manuscrit latin 4802 de la Bibliothèque nationale*, ed. Henri Auguste Omont (Paris: Catala Frères, 1926), where the *Egyptus novelo* map is on plates 64–65.

[11]Bertrand Hirsch, "Les sources de la cartographie occidentale de l'Éthiopie (1450–1550): les régions du lac Tana," *Bulletin des Études Africaines de l'INALCO* 7.13–14 (1987), pp. 203–236 and Bertrand Hirsch, "Connaissance et figures de l'Éthiopie dans la cartographie occidentale du XIVe siècle au XVIe siècle," Thèse de Doctorat, Université de Paris I, 1990.

Fig. 5.2 One of the three surviving Egyptus Novelo *maps, rotated so that north is at the top (Paris, Bibliothèque nationale de France, MS lat. 4802, ff. 130v–131r, made c. 1475–80). (By permission of the Bibliothèque nationale de France)*

draws valuable and indeed essential conclusions about the three surviving maps. In his 1987 article, he shows that none of the three surviving *Egyptus Novelo* maps can be considered the prototype or source of the other two: each of them contains toponyms that are absent from the others, as well as its own variants. The map in Vat. lat. 5699 is the most complete for Ethiopia, but the map in BnF MS lat. 4802 has significant details that are not present in the map in Vat. lat. 5699. He also supplies tables of some, but not all, of the toponyms on all three surviving versions of the map.[12] In another article he concludes that the map in BnF MS lat. 4802 is the most distant from the original or prototype of the map.[13]

In his unpublished 1990 dissertation, Hirsch offers a good discussion of the maps, emphasizing that Pietro del Massaio was the artisan who painted the maps rather than their creator, and noting that the map in Vat. lat 5699 is the richest in toponyms with 209, while that in BnF lat. 4802 has 191, and that in Vat. lat. 277 has 169, but again making it clear that all of the three maps have names which the others do not, which entails that they were independently copied from an older and more complete archetype. He also remarks on the lack of any explanatory text accompanying the map and speculates that there may have originally been such a text. In addition, he argues that multiple people from Ethiopia were questioned in order to obtain all of the 200 toponyms and other geographical data in the maps, noting that there is no equally large collection of geographic information about Ethiopia in any contemporary source, not even in the work of Pietro Rombulo da Messina, who lived for some 37 years in Ethiopia, of which excerpts are preserved in the manuscript *Annales omnium temporum* of Pietro Ranzano.[14] Hirsch also suggests that although there is no good evidence to determine the date of the (lost) prototype of the map, it seems most likely that the information presented in the map reached Europe at the Council of Florence in 1441.[15] In his dissertation Hirsch supplies complete tables of the toponyms on all three surviving versions of the

[12] See Hirsch, "Les sources de la cartographie occidentale de l'Éthiopie" (see note 11 in Chap. 5), esp. pp. 226 and 233–235.

[13] See Bertrand Hirsch, "Cartographie et itinéraires: Figures occidentales du nord de l'Ethiopie aux XVe et XVIe siècles," *Abbay* 13 (1986), pp. 91–122, at 107.

[14] See Hirsch, "Connaissance et figures de l'Éthiopie" (see note 11 in Chap. 5), vol. 1, pp. 174–188, titled "La *tabula moderna* de l'Ethiopie." On Pietro Rombulo see Carmelo Trasselli, "Un italiano in Etiopia nel XV secolo: Pietro Rombulo da Messina," *Rassegna di Studi Etiopici* 1 (1941), pp. 173–202; for a brief account in English, see O. G. S. Crawford, *Ethiopian Itineraries, ca. 1400–1524* (Cambridge: The Hakluyt Society, 1958), pp. 5–8; on Pietro Ranzano see Bruno Figliuolo, "Europa, Oriente, Mediterraneo nell'opera dell'umanista palermitano Pietro Ranzano," in Sergio Gensini, ed., *Europa e Mediterraneo tra Medioevo e prima età moderna: l'osservatorio italiano* (Pisa: Pacini, 1992), pp. 343–361 and Bruno Figliuolo, "L'umanista e teologo palmermitano Pietro Ranzano (1426/1427–1492/1493)," in *La cultura a Napoli nel secondo Quattrocento* (Udine: Forum, 1997), pp. 87–276, and on the *Annales omnium temporum,* see esp. 164–200. Ranzano's *Annales omnium temporum* are in Palermo, Biblioteca Communale, 3. Qq. C. 54–60.

[15] See Hirsch, "Connaissance et figures de l'Éthiopie" (see note 11 in Chap. 5), vol. 1, pp. 187–188. Hirsch also offers a general discussion of the transfer of geographical knowledge at the Council of Florence in his vol. 1, pp. 166–174, titled "La *Géographie*, les érudits et le concile de Florence."

Egyptus Novelo map,[16] in contrast to the incomplete tables in his article cited above.

Hirsch also argues that the river system in southern Africa on the Yale Martellus map shows the influence of the *Egyptus Novelo* but is not able to adduce any further evidence regarding the connection between these maps because he could not read any of the relevant toponyms. He suggests that Martellus saw the *Egyptus Novelo* map in a manuscript of Ptolemy's *Geography* illuminated by Pietro del Massaio and that the coastal contours of southern Africa on the Yale Martellus map come from Portuguese sources.[17] Having been able to read many of the African toponyms on the Yale Martellus map, I will be able to demonstrate Martellus's reliance on the *Egyptus Novelo*, but I will also be able to show that in all probability Martellus did not use a copy of the map in a manuscript painted by Pietro del Massaio. In fact Martellus had access to a more complete version of the geographical data than what is preserved in the surviving *Egyptus Novelo* maps, for on the Yale Martellus map what seems to be closely related data—the same river systems with accompanying named cities—extend beyond what is the eastern edge of the *Egyptus Novelo* maps. I will also take issue with Hirsch's claim that the contours of southeastern Africa, at least, come from Portuguese sources. In addition, I will supply many references regarding the toponyms on Martellus map (and on the *Egyptus Novelo* maps).

I now present the toponyms from the interior of southeastern Africa on the Yale Martellus map, beginning around the Tropic of Capricorn and proceeding from northwest to southeast. A number of the place names are legible in the c. 1960 ultraviolet photos of the map (see Fig. 2.7), but they are clearer in the 2014 multispectral images (tiles C04R05 and C05R05). Many of the toponyms from the *Egyptus Novelo* maps also appear on Martin Waldseemüller's world map of 1507 and his *Carta marina* of 1516, so some of my comments on Martellus's toponyms will involve those maps.

gâma. *The name appears beneath a city symbol. This toponym appears on the* Egyptus Novelo *maps, as* Gâma *in Vat. lat. 5699,* Gêma *in Urb. lat. 277, and* Garma *in BnF MS lat. 4208; on Waldseemüller's* Carta marina, *it is* Garma ecclesia *and is beside an image of a church.*[18]

huogi. *The name appears beneath a city symbol. The toponym is on all three* Egyptus Novelo *maps as* Huoggi *and on Waldseemüller's* Carta marina *as* Heuggi; *Randles lists it under* Euggi *and says that it is to be identified with* Ogge *or* Wagē,[19] *by which he means the former kingdom*

[16] See Hirsch, "Connaissance et figures de l'Éthiopie" (see note 11 in Chap. 5), vol. 2, pp. 17–26.

[17] See Hirsch, "Connaissance et figures de l'Éthiopie" (see note 11 in Chap. 5), vol. 1, pp. 274–284, titled "Henricus Martellus, inventeur d'une 'Afrique éthiopienne'."

[18] Hirsch does not list *Garma* in "Les sources de la cartographie occidentale de l'Éthiopie" (see note 11 in Chap. 5); he does list it in "Connaissance et figures de l'Éthiopie" (see note 11 in Chap. 5), vol. 2, p. 21, but does not indicate that it appears on Waldseemüller's *Carta marina*. It is included in the list of *Egyptus novelo* toponyms supplied by W. G. L. Randles, "South East Africa and the Empire of Monomotapa as Shown on Selected Printed Maps of the sixteenth Century," *Studia* 2 (1958), pp. 103–163, at 153. Randles on p. 121 says that Garma is unidentifiable.

[19] See Hirsch, "Connaissance et figures de l'Éthiopie" (see note 11 in Chap. 5), vol. 2, p. 22; and Randles, "South East Africa and the Empire of Monomotapa" (see note 18 in Chap. 5), pp. 121 and 153.

of Wäj, whose name was spelled in a bewildering number of variants but certainly included Lake Zway or Lake Ziway, southeast of Addis Ababa in modern Ethiopia.[20]

serbria. *This city is east of* gâma. *The name does not appear on any of the three* Egyptus Novelo *maps or on Waldseemüller's* Carta marina. *This name suggests that Martellus had access to a more complete version of the* Egyptus Novelo *map and that this name comes from that more complete version.*

Sedaia magna. *This city is just east of* huogi *(see above), and the corresponding city on all three* Egyptus Novelo *maps is* Sedaicion.[21] *On Martellus's map the city is at a fork in the* calicena fluvius, *while on the* Egyptus Novelo *maps,* Sedaicion *is near the* Hauasce fl., *so Martellus was working with a somewhat different version of the* Egyptus Novelo *map. On Waldseemüller's* Carta marina, *the city is called* Zedaia *and it is near the* Huilase fl.

bali. *This name is beside a city symbol a short distance southeast of* Sedaia. *The name does not appear on the* Egyptus Novelo *maps, but was an Ethiopian kingdom situated on the Shebelle River in Ethiopia, perhaps reaching as far south as the Gebele River and extending to the north of the Shebelle as well.*[22] *The presence of this Ethiopian place name on Martellus's map confirms that he was using a version of the* Egyptus Novelo *that was more complete than any of the three surviving exemplars. Bali does appear on Waldseemüller's* Carta marina, *but further to the north, well north of the Tropic of Capricorn, in contrast with its position south of the Tropic on the Yale Martellus map.*

vigt. *This city is on the Indian Ocean coast and thus is too far east to appear on the* Egyptus Novelo *maps—and is one of the few coastal cities that was apparently depicted on the more extensive version of the* Egyptus Novelo *map that Martellus was consulting.*

calicena fluvius. *There is no river of this name on the* Egyptus Novelo *or on Waldseemüller's* Carta marina. *As mentioned above, on the* Egyptus Novelo *maps,* Sedaicion *is on the* Hauasce fluvius, *while on Martellus it is on this river, which is more strong evidence that Martellus was working with a somewhat different map of Ethiopia.*

damat aurum. *This city is southwest of* huogi; *it is labeled* damot aurum *on all three of the surviving* Egyptus Novelo *maps, and* damot *on Waldseemüller's* Carta marina.[23] *This kingdom was south of the Abay River and west of the Muger River.*[24] *I do not know the basis for the toponym's apparent implication that the region produced gold.*

gafut. *This city is called* Gaffat *on all three* Egyptus Novelo *maps and* gafat *on Waldseemüller's* Carta marina. *On Martellus's map it is directly south of* damat aurum, *but on the* Egyptus Novelo *maps, it is directly west of* damot aurum, *a puzzling difference. Also, while on the*

[20] On the kingdom of Wäj, see Ulrich Braukämper, "Islamic Principalities in Southeast Ethiopia Between the Thirteenth and Sixteenth Centuries," *Ethiopianist Notes* 1.1 (Spring, 1977), pp. 17–56 and 1.2 (Fall, 1977), pp. 1–43, in part 1, pp. 41–47; these two parts of the article are reprinted together in Ulrich Braukämper, *Islamic History and Culture in Southern Ethiopia: Collected Essays* (Münster: Lit, 2002), pp. 12–105, in which the section on Wäj is on pp. 43–50.

[21] On *Sedaia magna* or *Sedaicion*, see Hirsch, "Connaissance et figures de l'Éthiopie" (see note 11 in Chap. 5), vol. 2, p. 24; Randles, "South East Africa and the Empire of Monomotapa" (see note 18 in Chap. 5), p. 155, lists the city under Zedaia.

[22] On Bali see Braukämper, "Islamic Principalities" (see note 20 in Chap. 5), part 2, pp. 21–29; and his *Islamic History and Culture in Southern Ethiopia* (see note 20 in Chap. 5), pp. 76–87; also see Richard Pankhurst, *The Ethiopian Borderlands: Essays in Regional History from Ancient Times to the End of the eighteenth Century* (Lawrenceville, N.J.: Red Sea Press, 1997), pp. 135–137 and 196–201. The toponym appears later (as Balli) on Giacomo Gastaldi's map of Africa: see Renato Biasutti, "La carta dell'Africa di G. Gastaldi (1545–1564) e lo sviluppo della cartografia africana nei sec. 16 e 17," *Bollettino della Reale Società Geografica Italiana*, Series 5, vol. 9 (1920), pp. 327–346 and 387–436, at 402.

[23] On *damot aurum* see Hirsch, "Connaissance et figures de l'Éthiopie" (see note 11 in Chap. 5), vol. 2, p. 19.

[24] For discussion of Damot, see C. Conti Rossini, "Geografica: 1. L'Africa orientale in carte arabe dei secoli XII e XIII; 2. Carte Abissine; 3. Gli itinerari di Alessandro Zorzi," *Rassegna di Studi Etiopici* 3 (1943), pp. 167–199, esp. 183, 185–186, 191, and 193; and Tsehai Berhane-Sellassie, "The Question of Damot and Walamo," *Journal of Ethiopian Studies* 13.1 (1975), pp. 37–45. There is an account of the area in the late nineteenth century in Gustavo Bianchi, "Damot," *L'Esploratore* 9 (1881), pp. 303–310.

Egyptus Novelo *maps Gaffat is near* Vabi lacus dulcis, *on Martellus there is no lake near the city. The toponym refers to the Gaffat people south of the Abay River, adjoining Damot.*[25]

uigt c m *(probably for* uigt civitas magna*). The city is east of* gafut. *This toponym does not appear on any of the three surviving* Egyptus Novelo *maps, but it does appear on Contarini's 1506 world map as* Vigict Magna *(which previously was thought to be the first occurrence of the toponym)—this is very strong additional evidence of the influence of Martellus on Contarini.*[26] *The name also appears on Waldseemüller's* Carta marina *as* vigiti magna, *directly south of* damot. *As there is not much room south of* Damot aurum *on the surviving* Egyptus Novelo *maps, perhaps this may be taken as evidence that Martellus and Waldseemüller had access to a version of the map that extended further south than the three surviving exemplars.*

hurab. *This city is south of* huogi. *The three exemplars of the* Egyptus Novelo *all give* Huorab, *and Waldseemüller on his* Carta marina *has* Huorob. *The toponym is evidently to be identified with Guerāgē or Gurage, a province in Abyssinia.*[27]

massara. *This city is south of* hurab *and is located on the* massara fluvius. *On the* Egyptus Novelo *map in BnF MS lat. 4802, the river has its name, and the toponym for the city appears on the river, but the dot that would indicate a city is missing, so that the toponym seems to be a repetition of the name of the river. The dot is present on the map in BAV Urb. lat. 277. On the* Egyptus Novelo *maps, the city is south of* Huorab, *as it is on Martellus's map. On Waldseemüller's* Carta marina *the city is* mossata *and is directly south of* Huorob. *The city of* masara *appears on Fra Mauro's* mappamundi[28]; *Crawford attempts to identify the relevant landmarks on Fra Mauro's map, but the matter is uncertain*[29]:

> [M]asara. A place near a lake (unnamed) given as the source of a southern (left-bank) tributary of the river Abaui. The lake and river may be the Jakjak and Fingar, respectively; or the lake may be the lake and swamp of Chomen higher up on the Fingar. But these identifications are doubtful, and the name *Masara* does not appear to survive there. *Egyptus Novelo* gives Massara as the name of a river rising in an unnamed lake and flowing into the Abaui (not named) south of Chiamo.

massara fl. *The same river appears on the* Egyptus Novelo *maps. Crawford identifies the river with the Muger, a tributary of the Abay River in Ethiopia.*[30]

---

[25] On *gafut* or *gaffat*, see Hirsch, "Connaissance et figures de l'Éthiopie" (see note 11 in Chap. 5), vol. 2, p. 21; on Gafat see Wolf Leslau, "A Short Chronicle of the Gafat," *Rivista degli studi orientali* 41 (1966), pp. 189–198; Randles, "South East Africa and the Empire of Monomotapa" (see note 18 in Chap. 5), p. 122; Braukämper, "Islamic Principalities in Southeast Ethiopia" (see note 20 in Chap. 5), part 2, p. 4, and Braukämper, *Islamic History and Culture in Southern Ethiopia* (see note 20 in Chap. 5), p. 52; and Pankhurst, *The Ethiopian Borderlands* (see note 22 in Chap. 5), pp. 89–90 and 260–264.

[26] For references on Contarini's map, see above note 82 in Chap. 1. For the claim that Contarini's was the first map on which this toponym appears see Randles, "South East Africa and the Empire of Monomotapa" (see note 18 in Chap. 5), pp. 109 and 154. Randles says that the name may be a corruption of the name of the river Wanshit in Abyssinia, but this seems dubious. There are some speculations about the source of the name Vigiti Magna, written without knowledge of the *Egyptus Novelo* map, and some remarks about it on later maps, in F. R. Paver, "A Puzzle in Antique Cartography: Vigiti Magna," *Africana Notes and News* 11.6 (1955), pp. 188–196.

[27] On *hurab* or *Huorab* see Hirsch, "Connaissance et figures de l'Éthiopie" (see note 11 in Chap. 5), vol. 2, p. 22; on the identification of the toponym, see W. G. L. Randles, "South-East Africa as Shown on Selected Printed Maps of the Sixteenth Century," *Imago Mundi* 13 (1956), pp. 69–88, esp. 85; Randles, "South East Africa and the Empire of Monomotapa" (see note 18 in Chap. 5), p. 121; Braukämper, "Islamic Principalities in Southeast Ethiopia" (see note 20 in Chap. 5), part 2, pp. 4–6; and Braukämper, *Islamic History and Culture in Southern Ethiopia* (see note 20 in Chap. 5), pp. 51–55.

[28] See Falchetta, *Fra Mauro's World Map* (see note 85 in Chap. 1), pp. 200–201, no. *91; Falchetta says that the city is unidentified.

[29] O. G. S. Crawford, *Ethiopian Itineraries*, ca. *1400–1524* (Cambridge: The Hakluyt Society, 1958), p. 200.

[30] On *massara fl.* see O. G. S. Crawford, "Some Medieval Theories about the Nile," *Geographical Journal* 114 (1949), pp. 6–23, at p. 12.

ualamo. *The city is just southeast of the city of* massara. *This corresponds to the city of* Vamo *on the map in BnF MS lat. 4802 (which is east of the city of* massara*), but the city does not appear on the other to* Egyptus Novelo *maps. Waldseemüller on his* Carta marina *has* Vallama *east of* mossata, *and this certainly corresponds to* Vamo*—the similarity of Waldseemüller's toponym to Martellus's, as well as the similarity of locations, confirm this.*

zayt mons. *This mountain is just south of* ualamo *and corresponds to a mountain named* zait *on the map in Vat. lat. 5699 and* Zau mons *on the map in Urb. lat. 277; the mountain is represented but unlabeled on the map in BnF MS lat. 4802.*[31] *On Waldseemüller's* Carta marina, *it is* Zaet mons*—see below. Randles notes that there was a province in Abyssinia named Zethe,*[32] *but the identification is far from certain.*

maia furzegar. *This city, which is directly north of* zayt mons, *certainly corresponds to* Maia fortes gentes *on the* Egyptus Novelo *maps, which is north of the corresponding mountain on those maps. Martellus's* furzegar *is puzzling; it is possible that the version of the* Egyptus Novelo *that he was consulting was difficult to read and that* furzegar *was his best guess for a poorly written* fortes gentes, *or perhaps we should consider* furzegar *the* lectio difficilior *and* fortes gentes *an attempt by a copyist to make sense of a strange name. On the other hand, the Maia were known as archers, so the designation* fortes gentes *is appropriate.*[33]

malza. c. *The city is east of* maia furzegar, *and the image of the city differs from the others examined thus far in that it bears two crosses. This toponym does not appear either on the* Egyptus Novelo *maps or on Waldseemüller's* Carta marina. *This city is thus more evidence that Martellus was using a different and more complete version of the* Egyptus Novelo *map as a source; it is unfortunate that it does not seem possible to identify one of the few Christian cities in the region.*

bali. *This toponym is beneath a city that is just northeast of* malza. c. *and just south of the Tropic of Capricorn. This name is repeated, apparently by mistake, beneath* gattiana civitas *(see the next entry). The name may be a repetition of* bali, *which was discussed above.*

gattiana civitas ubi moratur presbiter Johannnes. *"The city of Gattiana where Prester John lives." The city symbol just to the left of this text is much larger than all of the others in southern Africa on Martellus map—at least twice their size—and it has three crosses. The discovery of the city of Prester John in southern Africa on the Martellus map, particularly in an area where most of the other toponyms come from the* Egyptus Novelo, *is one of the most surprising and exciting I made in my study of the map, and the presence of the city merits a separate discussion, which I offer below.*

massara fluvius. *The river's name is repeated just below* zayt mons.

[banner]. *Just below* zayt mons *and the repetition of* massara fluvius, *there is a banner which is unfortunately illegible, even in the multispectral images. It would be extremely interesting to know what this banner says.*

barque fluvius. *There is no river of this name on the* Egyptus Novelo *maps. As the river does not reach the coast and is thus a purely interior feature, it seems likely that it comes from the larger version of the* Egyptus Novelo *map that Martellus was consulting. That larger version evidently extended further to the south and to the east of the area covered by the three* Egyptus Novelo *maps that have come down to us.*

elcha. *This city is on the* barque fluvius, *and there is nothing similar on the* Egyptus Novelo *maps.*

arna. *This city is on the southward continuation of the* calicena *river; again, it is not on any of the* Egyptus Novelo *maps or on Waldseemüller's* Carta marina.

manze. *This city is to the north, just east of Prester John's city.*

vagabariz. *This city is north of* manze *and is one of the very few coastal cities in this area.*

[31] These readings are from Hirsch, "Connaissance et figures de l'Éthiopie" (see note 11 in Chap. 5), vol. 2, p. 25.

[32] Randles, "South East Africa and the Empire of Monomotapa" (see note 18 in Chap. 5), pp. 117 and 155, citing Biasutti, "La carta dell'Africa di G. Gastaldi (1545–1564)" (see note 22 in Chap. 5), p. 402.

[33] On the Maia people, see Braukämper, "Islamic Principalities in Southeast Ethiopia" (see note 20 in Chap. 5), part 1, pp. 43–44; Braukämper, *Islamic History and Culture in Southern Ethiopia* (see note 20 in Chap. 5), pp. 45–46; and Pankhurst, *The Ethiopian Borderlands* (see note 22 in Chap. 5), p. 133.

baal. *This city is also to the north of* manze *and further to the east.*

falef. *The city is north of* baal, *near the Indian Ocean coast—one of the few cities near the coast in this area.*

alioar. *The city is southeast of* falef *and northeast of* baal.

provincia damoch. *This name is immediately south of* alioar *but only appears clearly on one multispectral image*[34]*: it seems that Martellus wrote the names of regions, including* provincia Mogo[] *below, in a different pigment than the names of cities and rivers. I have not found a source for this name. The name is reminiscent of* damat aurum *but that is much further to the west.*

ualla. *This city is just south of* baal, *on the southward continuation of the* calicena *river.*

mellae. *Just east of* ualla, *on the* calicena *river.*

alata. *Downstream of* mellae *on the* calicena *river.*

oarab. *In a small group of cities near the Indian Ocean coast.*

alcodor. *Closer to the coast than* oarab.

aleta. *Another city on the* calicena *river, near the Indian Ocean coast.*

manissa. *This city is southwest of* ualla, *on the* barque fluvius.

massara. *The river is named again, and this is important evidence: it shows that the same river that appears on the surviving* Egyptus Novelo *maps continued beyond the eastern boundary of those maps on the more complete version that Martellus was consulting.*

provintia Mogo[]. *This name is only visible in two multispectral images of the map,*[35] *which indicates that it was written in a different pigment than the other names nearby. The last part of the name is not legible; it might be* Mogom *or* Mogoia. *I have not managed to find a source for the name.*

barque fluvius. *The name of the river is repeated.*

delaneso civitas. *This city is the last on the western side of the barrier of mountains that runs north and south to separate the island or peninsula that forms the southeastern tip of Africa from the African continent proper. The central tower of the city symbol is taller than those on Martellus's other city symbols in the area, and it is topped by a cross. I have not been able to find any source for the name of the city, which was evidently thought to be the last Christian outpost before—what exactly?*

The barrier of mountains that runs north and south across the eastward-jutting peninsula here is where the *calicena*, *massara*, and *barque* rivers have their sources, and specifically, the *calicena* river rises from two lakes just west of the mountains that are reminiscent of the *paludes Nili* north of the Mountains of the Moon on Ptolemaic maps, from which the Nile flows northward.[36] The multispectral images (tile C05R05) reveal that the tip of the peninsula is separated from the rest of the continent not only by barrier of mountains but also by a channel that runs north and south just to the east of the mountains (Fig. 5.3). Moreover, the southeastern tip of the continent has mountains all along its northern and eastern coasts; we cannot see whether there are mountains along its southern coast, for that shore is covered by a banner that reads *GRADUS LONGITUDINIS AB*

[34]The image on which *provincia damoch* appears Supplementary Image 5.1 (C05R05_RF + FL_cal_r90_med3_stats_C03R03_PC7).

[35]The images in which *provincia Mogo* appears Supplementary Image 5.1 (C05R05_RF + FL_cal_r90_med3_stats_C03R03_PC7) and Supplementary Image 5.2 (C05R05_RF + FL_cal_r90_bands_05 + 06 + 12 + 16_lake_name_cyan__RIC1_GPC1_BIC1_hue-130_as_L_in_RGB).

[36]On the Renaissance cartography of these lakes, see Francesc Relaño, *Paludes Nili: la persistencia de las ideas ptolemaicas en la cartografía renacentista* (Barcelona: Facultad de Geografia i Historia, Cátedra de Geografía Humana, 1992) = *Geo-Crítica* 96.

Fig. 5.3 Multispectral image of the southeastern part of Africa on the Yale Martellus map (tile C05R05). (Image by Lazarus Project/MegaVision/RIT/EMEL, courtesy of the Beinecke Rare Book and Manuscript Library)

OCCIDENTE IN ORIENTEM. But aside from the question of the southern shore, the tip of the peninsula is formidably encircled.

In the western part of this enclosed area, there are three cities along the channel that runs north and south. Along the mountain range on the northeastern coast of the peninsula, a river flows to the southeast. There are three cities along this river, and the river ends in a lake with a label in it. The toponyms on the peninsula are difficult to read, even in the multispectral images (tile C05R05):

sador. *This is the city in the northwestern corner of the peninsula.*

telle. *This is the second city from the north along the channel that runs north and south.*

einrir[?]. *This is the third city from the north along the channel that runs north and south.*

laya. *This is the northernmost city near the northeastern shore of the peninsula; the city symbol and name are only visible in one multispectral image.*[37]

olm. *A city to the southeast of* laya *along the northeastern shore of the peninsula, between the branches of a river.*

cire[ ]. *The next city to the south, closer to the lake.*

elilus[?]. *The city closest to the lake.*

Lac[us] sarr [?]. *I have not been able to determine a likely source of this name.*[38]

---

[37]The image that reveals *Laya* is Supplementary Image 5.3 (C05R05_lp91bdPC2_transIC4i_statsIC2i_PS).

[38]The image that reveals *Lac*[*us*] *sarr* is Supplementary Image 5.4 (C05R05_RF FL_cal_r90_box101 med3_bad_b01-B12 B16 B18_14_bands_lake_name_roi_ICA4_cropped).

Unfortunately the names that are legible here provide no hint as to the source of the unusual and specific geography of the peninsula.

The toponyms in and configuration of southeastern Africa on Martellus map raise several issues that require discussion.

The first of these issues is the presence of the city of Prester John in southern Africa, with the legend *gattiana civitas ubi moratur presbiter Johannnes*, "The city of Gattiana where Prester John lives." Prester John was an essentially mythical Christian king thought to live somewhere in the East who possessed great wealth and power; stories about him were popular from the twelfth to the seventeenth century.[39] The name of Prester John's city, *gattiana*, is strange, but there is one other map that places a city with a somewhat similar name near Prester John's capital in southeastern Africa. On Giovanni Leardo's world map of 1448 now in Vicenza (Biblioteca Civica Bertoliana, MS 598 A),[40] in southern central Africa, there is a large symbol of a city that bears the legend *Impere de pro Joane*, and just south of that, there are four small squares indicating cities, and one of these is labeled *Grasati*.[41] It is perhaps tempting to see this similarity of name as coincidence, but then Leardo's 1448 map does have other similarities with the Yale Martellus map in Africa: it also has Prester John in the central part of the continent and a large peninsula jutting eastward from the southeastern part of the continent.

In and of itself, the presence of Prester John in Africa on the Yale Martellus map is not surprising: although Prester John was originally thought to be in Asia, his legend had been migrating to Africa for some time.[42] He first appeared in Africa on Angelino Dulcert's nautical chart of

[39]On Prester John see Charles F. Beckingham and Bernard Hamilton, eds., *Prester John, the Mongols, and the Ten Lost Tribes* (Aldershot, Hampshire; and Brookfield, VT: Variorum, 1996); and Michael E. Brooks, "Prester John: A Reexamination and Compendium of the Mythical Figure Who Helped Spark European Expansion," Ph.D. Dissertation, University of Toledo, 2009. The forged letter allegedly from Prester John that was circulated across Europe in the twelfth century is translated into English in Michael Uebel, *Ecstatic Transformation: On the Uses of Alterity in the Middle Ages* (New York: Palgrave Macmillan, 2005), pp. 155–160.

[40]For discussion of and transcription of all of the toponyms and legends on Leardo's 1448 map, see Santarém, *Essai sur l'histoire de la cosmographie et de la cartographie pendant le Moyen-âge* (Paris: Maulde et Renou, 1849–52), vol. 3, pp. 398–442; for a more recent discussion of the map with a good color illustration, see Cavallo, *Cristoforo Colombo e l'apertura degli spazi* (see note 42 in Chap. 1), vol. 1, pp. 159–162. The map is also illustrated in Santarém's *Atlas composé de mappemondes, de portulans et de cartes hydrographiques et historiques, depuis le VIe jusqu'au XVIIe siècle* (Paris: E. Thunot, 1849), and that image is reproduced in A. E. Nordenskiöld, *Periplus: An Essay on the Early History of Charts and Sailing-Directions* (Stockholm: P.A. Norstedt & söner, 1897), p. 61. The map is also said to be very well reproduced in this book, but I have not been able to see a copy of it: Mario Guderzo, *Teatro del cielo e della terra: mappamondi, carte nautiche e atlanti della Biblioteca bertoliana dal XV al XVIII secolo* (Vicenza: Biblioteca Civica Bertoliana, 1984).

[41]The toponym *Grasati* is listed by Santarém, *Essai sur l'histoire de la cosmographie* (see note 40 in Chap. 5), vol. 3, p. 436.

[42]On the transfer of Prester John from Asia to Africa see J. Richard, "L'Extrême-Orient légendaire au Moyen Âge: Roi David et Prêtre Jean," *Annales d'Éthiopie* 12 (1957), pp. 225–244; Jeremy Lawrance, "The Middle Indies: Damião de Góis on Prester John and the Ethiopians," *Renaissance Studies* 6.3–4 (1992), pp. 306–324; Relaño, *The Shaping of Africa* (see note 43 in Chap. 1), pp. 51–72; and Manuel João Ramos, "Rex, Sacerdos et Niger," in his *Essays in Christian Mythology: The Metamorphosis of Prester John* (Lanham, MD: University Press of America, 2006), pp. 106–116.

1339,[43] and there was a mid-fifteenth-century nautical chart (now lost) whose legends are preserved which offers an account of Prester John's transfer from Asia to Africa.[44] But there are some aspects of his presence in Africa on the Yale Martellus map that are of particular interest. First, although Prester John does not appear on the world maps in the manuscripts of Martellus's *Insularium* in Florence or Leiden nor on the Martellus-Rosselli map, he does appear on the world map in London—but in Asia rather than in Africa (see Fig. 1.2). Moreover, on Waldseemüller's 1507 map, Prester John is in Asia in almost exactly the same spot as on the map in the London manuscript, so in this matter Waldseemüller certainly did not follow the Yale Martellus map. This difference between the map in the London manuscript and the Yale Martellus map confirms that Martellus was following quite a different source when he made his Yale map and also that Waldseemüller was using a version of Martellus large map different from that at Yale.

Second, it should be emphasized that it is quite possible that the information about the city of Prester John on the Yale Martellus came from an Ethiopian source, just like the rest of the data in this part of the map. For the Ethiopian delegates at the Council of Florence (1439–45), when asked about Prester John, readily replied that he was a sovereign in Ethiopia[45]:

> *De rege interrogati, cui apud nos et in Syria ac Aegypto presbytero Iohanni est appellatio, responderunt: eum, qui indigne ferat hanc absurdam sibi natam appellationem, nomine proprio Zareiacob, quod interpretari volunt Iacob prophetae soboles, sed cognomine et simul dignitate Constantinum dici, tamquam virtute et potentia illi similem, qui primus Romani imperii sedem transtulit Byzantium, Flavio Constantino: cui suo regi subesse affirmant centum reges maiora singulos possidentes regna quam nostri Europae, quorum diligenter narratum intelligerent potentatum.*
>
> *When asked about the king who among us and in Egypt and Syria is called Prester John, they answered that the king, who resents that name that he has acquired, is actually named Zareiacob, which name they wish to be interpreted as "offspring of the prophet Jacob," but he is called "Constantine" both for his surname and for his rank, since in both goodness and power he is similar to Flavius Constantinus, who first transferred the seat of Roman power to Byzantium. To this king of theirs, they assert that a hundred kings are subject, each of them possessing a realm that is larger than Europe, whose carefully described power they understand.*

So it is in fact quite probable that Prester John would appear on the larger and more detailed version of the *Egyptus Novelo* map that Martellus

[43] Angelino Dulcert's 1339 chart is in Paris, Bibliothèque nationale de France, Rés. Ge B 696; it is reproduced in Cavallo, *Cristoforo Colombo e l'apertura degli spazi* (see note 42 in Chap. 1), vol. 1, pp. 164–165, with descriptive text on pp. 162–163; it is reproduced on a larger scale in Gabriel Marcel, *Choix de cartes et de mappemondes des XIVe et XVe siècles* (Paris: E. Leroux, 1896); and in Ramon J. Pujades i Bataller, *Les cartes portolanes: la representació medieval d'una mar solcada* (Barcelona: Institut Cartogràfic de Catalunya, 2007), pp. 120–121 and on the accompanying CD, number C8.

[44] See Jacques Paviot, "Une mappemonde génoise disparue de la fin du XIVe siècle," in Gaston Duchet-Suchaux, ed., *L'Iconographie: études sur les rapports entre textes et images dans l'Occident médiéval* (Paris: Le Léopard d'Or, 2001), pp. 69–97, at 87–88 and 96. Paviot claims that the lost chart was from the late fourteenth century, but as one of its legends cites Antoniotto Usodimare, it must be from 1455 or later.

[45] The text is from Blondi Flavii Forliviensis, *Historiarum ab inclinatone Romanorum, Quartae Decadis Liber Secundus*, in Bartolomeo Nogara, ed., *Scritti inediti e rari di Biondo Flavio* (Rome: Tipografia Poliglotta Vaticana, 1927), p. 24. The translation is mine.

was using: the data from that map most likely came from the Ethiopian delegates at the Council of Florence, and as the passage just cited shows, those delegates knew of a sovereign in southern Africa called Prester John. At the same time, there are other European maps that also have a peninsula jutting eastward from southeastern Africa and Prester John in southern central Africa, including Giovanni Leardo's *mappaemundi* of 1448[46] and 1452 or 1453,[47] Fra Mauro's map of c. 1450,[48] the so-called Genoese map of 1457,[49] and two maps associated with Benedetto Cotrugli's *De navigatione*.[50]

The topography of southeastern Africa on the Yale Martellus map is similar to that of a place sometimes associated with Prester John, namely, the Terrestrial Paradise.[51] The whole peninsula is basically surrounded by

[46]Leardo's *mappamundi* of 1448 is in Vicenza, Biblioteca Civica Bertoliana, MS 598 A. For discussion of and transcription of all of the map's toponyms and legends see Santarém, *Essai sur l'histoire de la cosmographie* (see note 40 in Chap. 5), vol. 3, pp. 398–442; for a more recent discussion of the map with a good color illustration, see Cavallo, *Cristoforo Colombo e l'apertura degli spazi* (see note 42 in Chap. 1), vol. 1, pp. 159–162.

[47]Leardo's *mappamundi* of 1452 or 1453 is in Milwaukee, American Geographical Society Library, 050 A-1452. For discussion of it see Guglielmo Berchet, *Il Planisfero di Giovanni Leardo dell'anno 1452: fac-simile nella grandezza dell'originale* (Venice: Ferdinando Ongania Edit., 1880); and John Kirtland Wright, *The Leardo Map of the World, 1452 or 1453, in the Collections of the American Geographical Society* (New York: American Geographical Society, 1928).

[48]For Fra Mauro's legend about Prester John in Africa, see Falchetta, *Fra Mauro's World Map* (see note 85 in Chap. 1), pp. 194–195, no. *66.

[49]The "Genoese" world map of 1457 is in Florence, Biblioteca Nazionale Centrale, Portolano 1. For discussion of it, see Edward Luther Stevenson, *Genoese World Map, 1457* (New York: American Geographical Society and Hispanic Society of America, 1912), which was accompanied by a color facsimile of the map; Destombes, *Mappemondes* (see note 10 in Front Matter), pp. 222–223 and plate 34; Angelo Cattaneo, *Mappa mundi 1457* (Rome: Treccani, 2008), the study that accompanies a color facsimile of the map. It is conveniently reproduced in color in Cavallo, *Cristoforo Colombo e l'apertura degli spazi* (see note 42 in Chap. 1), vol. 1, pp. 492–493. The legends on the map are transcribed and translated by Stevenson and Cattaneo, and also in Sebastiano Crinò, *La scoperta della carta originale di Paolo dal Pozzo Toscanelli che servì di guida a Cristoforo Colombo per il viaggio verso il Nuova mondo* (Florence: Istituto geografico militare, 1941), pp. 28–32—though Crinò's attribution of the map to Toscanelli is untenable.

[50]The two maps associated with Cotrugli are Vatican City, Biblioteca Apostolica Vaticana, MS Ottob. lat. 1417, f. 2v; and Berlin, Staatsbibliothek, Preussischer Kulturbesitz, MS Hamilton 108, f. 81r. On the map in the Vatican, there is a legend in Africa that reads *India preti ianis*, and on the map in Berlin, there is a legend in Africa that reads *India presbyteris Joh*[*annis*]. See my article "Benedetto Cotrugli's Lost Mappamundi Found—Three Times," *Imago Mundi* 65.1 (2013), pp. 1–14.

[51]The *Letter of Prester John* mentions a river full of precious stones that flows from Paradise through a province of his realm, and also a precursor of the Fountain of Youth in one of his provinces, that is found just three days' journey from Paradise. The Latin text of the Letter is supplied by Friedrich Zarncke, "Der Priester Johannes," *Abhandlungen der philologisch-historischen Classe der Koeniglichen Sachsischen Gesellschaft der Wissenschaften* 7.8 (1879), pp. 826–1028 and 8.1 (1883), pp. 1–183, at 909–924; an English translation is supplied by Michael Uebel, *Ecstatic Transformation: On the Uses of Alterity in the Middle Ages* (New York: Palgrave Macmillan, 2005), pp. 155–160, where the passages on Paradise are on p. 156, sections 22 and 27. This association between Prester John and the Terrestrial Paradise is reflected, for example, in the late fourteenth-century *Itinerarius* of Johannes Witte de Hese, in which the palace of Prester John is said to be built over one of the rivers of Paradise, and the three other rivers are said to be nearby (267–268 and 293–297): see Scott D. Westrem, *Broader Horizons: A Study of Johannes Witte de Hese's Itinerarius and Medieval Travel Narratives* (Cambridge, MA: Medieval Academy of America, 2001), pp. 8, 141, 143, 219, and 220.

mountains, and there are maps in which the Terrestrial Paradise is surrounded by mountains. On the *mappamundi* in the late eleventh-century Saint-Sever manuscript of Beatus of Liébana's *Commentary on the Apocalypse*, it is completely surrounded by mountains[52]; on the Borgia world map of c. 1430, it is partly surrounded by mountains[53]; on the Catalan-Estense *mappamundi* of c. 1460, it is partly surrounded by flaming mountains which are said to be made of diamonds[54]; and on the *mappamundi* on the neck of the so-called Columbus Chart of c. 1492, it is completely surrounded by mountains.[55] On other maps Paradise is surrounded by a barrier that probably represents a wall, for example, on the Psalter *mappamundi* of c. 1265 and on the world map by Albertin de Virga of 1411 or 1415 (Fig. 5.4).[56]

Moreover, it is tempting to connect the lake in the peninsula, the enigmatically named *Lac[us] sarr*, with the fountain mentioned in Genesis 2:6 that features in many medieval depictions of Eden.[57] In addition, the three large rivers have their origin just outside the barrier of mountains that separates the peninsula from the rest of Africa are very similar indeed to the four rivers that are usually depicted issuing from the barrier surrounding the Terrestrial Paradise in medieval and Renaissance depictions of Paradise, including on maps. The presence of a prominent Christian city

[52]The Saint-Sever Beatus *mappamundi* is in Paris, Bibliothèque nationale de France, MS lat. 8878, ff. 45 bisv-45 terr; the manuscript has been reproduced in facsimile as Beatus of Liébana, *Comentario al Apocalipsis y al Libro de Daniel* (Madrid: Edilán, 1984).

[53]The Borgia metal *mappamundi* is in Vatican City, Biblioteca Apostolica Vaticana, Borgia XVI; the image of Terrestrial Paradise on the map is reproduced in Alessandro Scafi, *Mapping Paradise: A History of Heaven on Earth* (Chicago: University of Chicago, 2006), p. 212, Fig. 8.8, with discussion and supplementary illustrations on pp. 210–212. The map is briefly described by Destombes, *Mappemondes* (see note 10 in Front Matter), pp. 239–240, no. 53.1; all of the map's legends are transcribed in N. A. E. Nordenskiöld, "Om ett aftryck från XV:de seklet af den i metall graverade världskarta, som förvarats i kardinal Stephan Borgias museum i Velletri, Med 1 facsimile," *Ymer* 11 (1891), pp. 83–92, with the reproduction of the map between pp. 130 and 131. For a good detailed study of the map, see John Hamer, "The Borgia Map: Europe's Rise and the Re-Definition of the World," MA Thesis, University of Michigan, Ann Arbor, 1995.

[54]The Catalan Estense map is in Modena, Biblioteca Estense Universitaria, C. G. A. 1, and has been reproduced in facsimile, with transcription and commentary, in Ernesto Milano and Annalisa Battini, *Mapamundi Catalán Estense, escuela cartográfica mallorquina* (Barcelona: M. Moleiro, 1996); there is a high-resolution digital image of the map on the CD-ROM titled *Antichi planisferi e portolani: Modena, Biblioteca Estense Universitaria* (Modena: Il Bulino; and Milan: Y. Press, 2004), and a good study of it in Konrad Kretschmer, "Die katalanische Weltkarte der Biblioteca Estense zu Modena," *Zeitschrift der Gesellschaft für Erdkunde zu Berlin* 32 (1897), pp. 65–111 and 191–218. For discussion of Paradise on the map, see Scafi, *Mapping Paradise* (see note 53 in Chap. 5), pp. 226–230.

[55]The "Columbus Chart" is in Paris, Bibliothèque nationale de France, Rés. Ge AA 562; the *mappamundi* on the chart is conveniently reproduced in Nebenzahl, *Atlas of Columbus* (see note 42 in Chap. 1), p. 22; and in Scafi, *Mapping Paradise* (see note 53 in Chap. 5), plate 14 and figs. 8.12a and 8.12b, with discussion on p. 217.

[56]See Scafi, *Mapping Paradise* (see note 53 in Chap. 5), pp. 148–149 and plate 8 on the Psalter *mappamundi*, and pp. 219–222 on Albertin de Virga's world map.

[57]The fountain in Eden is mentioned in Genesis 2:6. See Paul A. Underwood, "The Fountain of Life in Manuscripts of the Gospels," *Dumbarton Oaks Papers* 5 (1950), pp. 41–138; Tania Velmans, "Quelques versions rares du thème de la fontaine de vie dans l'art paléochrétien," *Cahiers archéologiques* 19 (1969), pp. 29–43; Ernst Schlee, *Die Ikonographie der Paradiesesflüsse* (Leipzig: Dieterich'sche, 1937); and Marie-Thérèse Gousset, *Éden: le jardín médiéval à travers l'enluminure, XIIIe-XVIe siècle* (Paris: Albin Michel and Bibliothèque nationale de France, 2001).

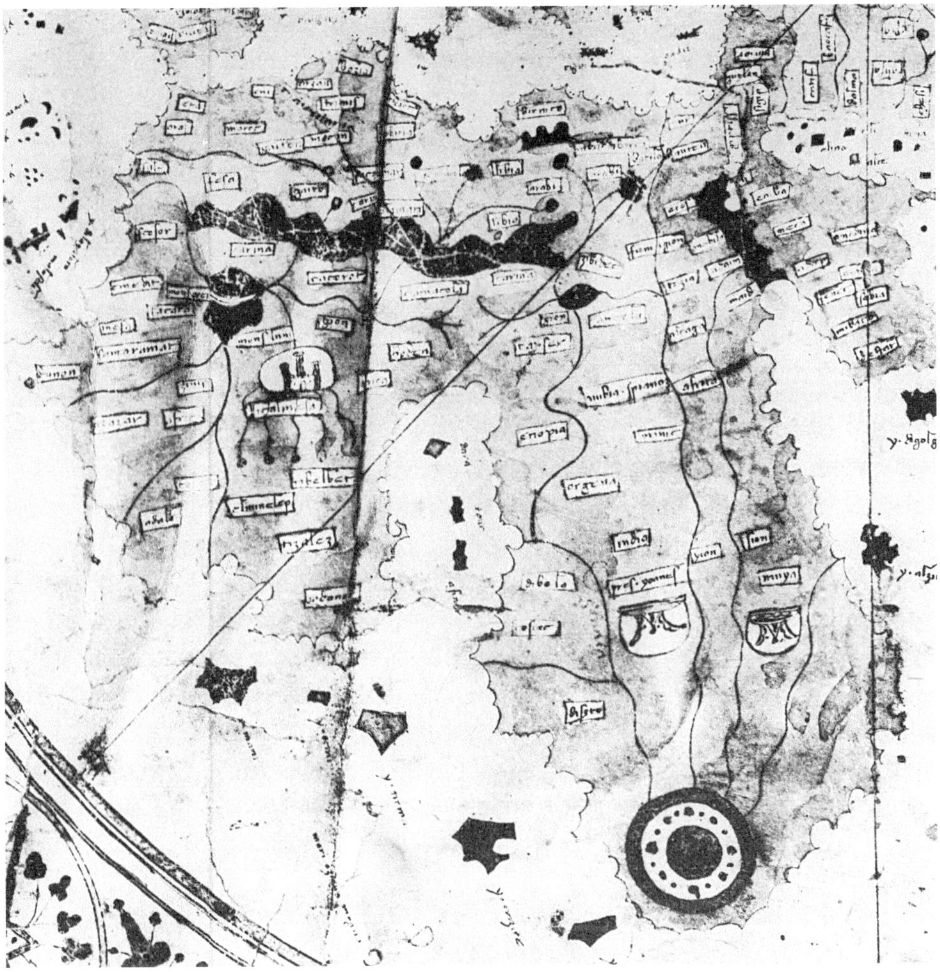

Fig. 5.4 Africa on Albertin de Virga's world map of 1411 or 1415, showing Paradise in the southern tip of the continent, from Yusuf Kamal, Monumenta cartographica Africae et Aegypti *(Cairo, 1926–51). Geography and Map Division, G2445 .Y8 folio. (Courtesy of the Library of Congress)*

just outside of the barrier of mountains also suggests that what is behind the mountains is of importance in the Christian religion. And the location of the peninsula, in southern Africa, is one that is occupied by the Terrestrial Paradise in other maps. These maps include the *mappamundi* in the *Llibre Vermell*, which dates from the middle of the fourteenth century; has a prominent representation of the Terrestrial Paradise in the southern part of the world, right on the line that separates Africa and Asia, and includes nearby the legend *preste iohan*.[58] Albertin de Virga's map of 1411 or 1414 also has the Terrestrial Paradise in southern Africa, with the legend *pres.*

[58]The *mappamundi* in the *Llibre Vermell* is Monistrol de Montserrat (Spain), Abadia de Montserrat, MS 1, f. 68r. There is a partial facsimile of the manuscript, whose introduction includes a detailed description and discussion of the manuscript, Francesc Xavier Altés i Aguiló, *Llibre Vermell de Montserrat: edició facsímil parcial del manuscrit núm. 1 de la Biblioteca de l'Abadia de Montserrat* (Barcelona: Fundació Revista de Catalunya, 1989), and this facsimile can be consulted online at http://www.lluisvives.com/servlet/SirveObras/jlv/08140629733581728654480/. Also see Cebrià Baraut, "Els manuscrits de l'antiga biblioteca del monestir de Montserrat (segles XI–XVIII)," *Analecta Montserratensia* 8 (1954–1955), pp. 339–396, esp. 342–348 on the *Llibre Vermell*, and 346 on the description of the universe that includes the *mappamundi*; and Pere Bohigas, "L'agrupament de les miniaturas del Llibre Vermell," *Analecta Montserratensia* 9 (1962), pp. 39–53, esp. 40 on the *mappamundi*, with a reproduction in Fig. 5.

jonnes just north of it (Fig. 5.4); and the Catalan Estense *mappamundi* of c. 1460[59] has Paradise in the easternmost reaches of Africa, and *presta johan* is depicted not far to the west.[60] Also the so-called Genoese map of 1457 includes a legend on its peninsula in southeastern Africa that says that some cosmographers placed Paradise in that location,[61] and Prester John is depicted not far to the northwest.

Thus it seems very likely that for his depiction of southeastern Africa, Martellus relied on the work of a cartographer who had located the Terrestrial Paradise in a peninsula or island at the southeastern tip of the continent. However, either the cartographer of that source map or Martellus himself altered things so as to efface any explicit reference to the traditional topography of the Terrestrial Paradise. For example, the lake in the eastern part of the peninsula seems to receive a river rather than be the source of it, and the rivers flowing from the western side of the barrier of mountains do not have the traditional names of the rivers of Paradise (Tigris, Euphrates, Gihon, and Phison, see Genesis 2:10–14) but rather the names of the rivers from the *Egyptus Novelo* map. Given that the two other maps on which the geography of Paradise is the most similar to that of the Yale Martellus map, namely, the so-called Genoese map of 1457 and the Catalan Estense *mappamundi*, date from the mid-fifteenth century, it is very likely that the source from which Martellus was working had its basis on a map from that period, though perhaps he was working from a later copy in which the toponymy related to Paradise had been modified or removed.

At the same time, the possibility of Ethiopian influence on Martellus's peninsula or island in southeastern Africa must also be considered. First, although the *Egyptus Novelo* map does not show the southeasternmost part of the continent, it does have a total of eight rivers flowing either to or from (it is difficult to tell which) the eastern part of the continent, so it seems certain that the cartographer believed there was a hydrographical

[59]The Catalan Estense map is in Modena, Biblioteca Estense Universitaria, C. G. A. 1, and has been reproduced in facsimile, with transcription and commentary, in Ernesto Milano and Annalisa Battini, *Mapamundi Catalán Estense, escuela cartográfica mallorquina* (Barcelona: M. Moleiro, 1996); there is a high-resolution digital image of the map on the CD-ROM titled *Antichi planisferi e portolani: Modena, Biblioteca Estense Universitaria* (Modena: Il Bulino; and Milan: Y. Press, 2004), and a good study of it in Konrad Kretschmer, "Die katalanische Weltkarte der Biblioteca Estense zu Modena," *Zeitschrift der Gesellschaft für Erdkunde zu Berlin* 32 (1897), pp. 65–111 and 191–218. The map is conveniently reproduced in color in Cavallo, *Cristoforo Colombo e l'apertura degli spazi* (see note 42 in Chap. 1), vol. 1, pp. 496–497.

[60]For discussions of Paradise on the Albertin de Virga map and the Catalan Estense *mappamundi*, see Scafi, *Mapping Paradise* (see note 53 in Chap. 5), pp. 219–222 and 226–230; and Francesc Relaño, "Paradise in Africa: The History of a Geographical Myth from its Origins in Medieval Thought to its Gradual Demise in Early Modern Europe," *Terrae Incognitae* 36 (2004), pp. 1–11.

[61]Here is the legend on the Genoese map; the Latin is from Stevenson, *Genoese World Map, 1457* (see note 49 in Chap. 5), p. 63, and the English translation is based on his: *In hac regione depinxerunt quidem paradisum deliciarum. Alii vero ultra Indias ad orientem eum esse dixerunt. Sed quoniam hec est cosmographorum descriptio qui nullam de eo fecerunt mentionem, ideo omititur hic de eo narratio*, that is, "In this region some have depicted the Terrestrial Paradise. But others have said that it is beyond the Indies to the east. But since that is a representation of cosmographers who have given no description of it, therefore an account of it is here omitted."

feature of great importance east of the current edge of the map, perhaps something analogous to the source of the world's great rivers believed to be in the Terrestrial Paradise. In addition—and this is very important—the eastern edge of the *Egyptus Novelo* map runs through the Red Sea, and so the eight rivers flowing in the area beyond the eastern edge of the map clearly indicate a substantial extension of land eastward into the Indian Ocean: the *Egyptus Novelo* maps that have come down to us clearly imply the existence of a substantial peninsula jutting eastward into the Indian Ocean, at least vaguely similar to what we see on the Yale Martellus map.

Further, there is a striking similarity between the outlines of southeastern Africa on the Yale Martellus map and Fra Mauro's *mappamundi* (Fig 5.5)[62]: both maps have a peninsula or island jutting eastward from the southeastern tip of the continent, and in both cases there is a barrier of mountains separating the island or peninsula from Africa proper.[63] Moreover, in both there is a lake in the eastern part of the island or peninsula with a river entering or leaving the lake to the northwest. There are also differences between the representations of these regions on the two maps: the island of Diab on Fra Mauro's map is not surrounded by mountains like the island or peninsula on Martellus map, and the Nile system does not reach eastward toward Diab as it does toward the island or peninsula on Martellus map. But the similarity of outline is undeniable, and Fra Mauro says explicitly that his depiction of southeastern Africa is based on a map that was made for him by Ethiopian clerics who evidently visited him in Venice[64]:

> *Perché ad alguni par da nuovo che io parli de questa parte meridional, la qual quasi està incognita a li antichi, perhò io respondo che tuto questo desegno da sayto in suso io l'ò habuto da queli proprij che sono nasudi qui, che sono stà religiosi, i qual cum le suo man me hano desegnato tute queste provincie e citade e fiumi e monti cum li suo nomi, le qual tute cose non le ho possudo meter cum el debito ordine per non esservi logo.*
>
> *Because to some it will appear as a novelty that I should speak of these southern parts, which were almost unknown to the Ancients, I will reply that this entire drawing, from Sayto upwards, I have had from those who were born there. These people are clerics who, with their own hands, drew for me these provinces and cities and rivers and mountains with their names; all these things I have not been able to put in due order for lack of space.*

[62] For transcription and analysis of all of the place names and legends on Fra Mauro's *mappamundi*, see Falchetta, *Fra Mauro's World Map* (see note 85 in Chap. 1); for discussion of the intellectual background of the map, see Angelo Cattaneo, *Fra Mauro's Mappa Mundi and Fifteenth-Century Venice* (Turnhout: Brepols, 2011).

[63] On Fra Mauro's map, the island is called Diab; on the channel that separates Diab from the mainland, see Falchetta, *Fra Mauro's World Map*, p. 191, *49; and for discussion of Diab, see Falchetta, *Fra Mauro's World Map* (see note 85 in Chap. 1), pp. 97–100, and p. 178, *18; Cerulli, Enrico, "Fonti arabe del 'mappamondo' di Fra Mauro," *Orientalia: commentarii periodici Pontifici Instituti Biblici* 4 (1935), pp. 336–338; and Bertrand Hirsch, "Connaissance et figures de l'Éthiopie dans la cartographie occidentale du XIVe siècle au XVIe siècle," Ph.D. Dissertation, Université de Paris I, 1990, vol. 1, pp. 210–220. Also see Renato Lefevre, "Riflessi etiopici nella cultura europea del Medioevo e del Rinascimento," *Annali Lateranensi* 8 (1944), pp. 9–89; 9 (1945), pp. 331–444, and 11 (1947), pp. 255–342, esp. 9 (1945), pp. 351–358.

[64] The text and translation of Fra Mauro are from Falchetta, *Fra Mauro's World Map* (see note 85 in Chap. 1), pp. 200–203, no. *98.

Fig. 5.5 Southeastern Africa on Fra Mauro's mappamundi *of c. 1455, Biblioteca Nazionale Marciana, Venice. By concession of the Ministero dei Beni e delle Attività Culturali e del Turismo – Biblioteca Nazionale Marciana. (Further reproduction is prohibited)*

Hirsch notes that he has not been able to identify the Ethiopian pilgrims with whom Fra Mauro claims to have spoken and says that there is no evidence that the Ethiopian delegates who attended the Council of Florence in 1441—who were probably the source of the information that appears in the *Egyptus Novelo* map—also visited Venice.[65] But in fact the Ethiopia of Fra Mauro's map shares just a few toponyms with the *Egyptus Novelo*,[66] so it would be difficult to believe that the same Ethiopians supplied the information for both maps. Moreover, interchanges between Ethiopia and Europe were more frequent during the fifteenth century than one might expect,[67] so it is perfectly possible that in addition to the Ethiopian delegates to the Council of Florence in 1441, Ethiopian clerics visited Venice during the years that Fra Mauro was researching and making his *mappamundi*.

What then was the nature of the map that Martellus was using as his source for southern Africa? It would seem to have been a version of the *Egyptus Novelo* map that extended beyond the eastern and southern edges of the three *Egyptus Novelo* maps that have come down to us, to include a substantial peninsula extending eastward into the Indian Ocean, which peninsula contained a significant hydrographical feature. Unfortunately there does not seem to be a way to choose among the various ways in which the Terrestrial Paradise ended up on Martellus map. Possible means of transmission include the following, but there are many other conceivable permutations:

1. The larger version of the *Egyptus Novelo* map, which was made through an interaction between Ethiopians and Europeans, contained the Terrestrial Paradise, and Martellus decided to accept the contours of the map, but did not believe that Eden had a physical location and so changed or did not copy the relevant toponyms.
2. Martellus was using a copy of the larger *Egyptus Novelo* map made by a European cartographer who had interpreted the large eastward-jutting peninsula as the Terrestrial Paradise and had made some changes to reflect this (such as adding mountains all the way round the peninsula),

[65] See Hirsch, "Connaissance et figures de l'Éthiopie" (see note 11 in Chap. 5), vol. 1, pp. 214–215.

[66] For toponyms from Fra Mauro's map that also appear on the *Egyptus novelo* see Falchetta, *Fra Mauro's World Map* (see note 85 in Chap. 1), pp. 194–195, no. *65, and pp. 202–203, no. *101, *Provincia fatagar*, which perhaps corresponds to *Façagur* on the *Egyptus novelo*; pp. 200 201, no. *91, *Masara*; pp. 278–279, no. 467, *Flumen mana*; and pp. 354–355, no. *852, *Aidab*.

[67] On fifteenth-century interactions between Ethiopia and Europe, see Renato Lefevre, "Riflessi etiopici nella cultura europea del Medioevo e del Rinascimento," *Annali Lateranensi* 8 (1944), pp. 9–89; 9 (1945), pp. 331–444; and 11 (1947), pp. 255–342; Taddesse Tamrat, "Early Contacts with Christian Europe (1380–1460)," in *Church and State in Ethiopia, 1270–1527* (Oxford: Clarendon Press, 1972), pp. 248–267; C. F. Beckingham, "Ethiopia and Europe, 1200–1650," in Cecil H. Clough and P. E. H. Hair, eds., *The European Outthrust and Encounter: The First Phase c.1400–c.1700: Essays in Tribute to David Beers Quinn on his 85th Birthday* (Liverpool: Liverpool University Press, 1994), pp. 77–95; Gianfranco Fiaccadori, "Italy, Relations With," in Siegbert Uhlig, ed., *Encyclopaedia Aethiopica* (Wiesbaden: Harrassowitz, 2003), vol. 3, pp. 236–239; and Kate Lowe, "'Representing' Africa: Ambassadors and Princes from Christian Africa to Renaissance Italy and Portugal, 1402–1608," *Transactions of the Royal Historical Society* 17 (2007), pp. 101–128.

and Martellus decided to accept the contours of the map, but did not believe that Eden had a physical location and so changed or did not copy the relevant toponyms.

3. The map that Martellus was using was similar to his: it had the toponyms and outlines of the larger *Egyptus Novelo* map, but the depiction of the Terrestrial Paradise already lacked some of the features and toponyms typical of European representations.

In fact there is a specific map, now lost, that may have been Martellus's source. In the Cedole della Tesoreria Aragonese, or records of the treasury of the Aragonese kings of Naples, which had been stored in the Archivio di Stato di Napoli but were burned in September of 1943, there was a record of a payment made by the treasurer for a large-format manuscript of Ptolemy's *Geography* and for a cloth on which was painted a map of Ethiopia, acquired on the behalf of Alfonso V of Aragon, who was King of Naples between 1443 and 1458.[68] Fortunately this record was copied before the burning of the records[69]:

> *1456, luglio.-- Item doni an Tomas Actani, mercader florentin qui atura de present en la ciutat de Napols, CCVI d. VIIII – los quals li eren deguts ab dos albarans de scrivan de racion en la forma seguent: ço es, ab lo primer dels dits albarans scrit en Napols a XXVIII dies del prope passat mes de marc, per lo preu de un libre appelat "Cosmografia Tolomei" de forma maior, scrit en pergamins de letra antiga, en lo principi del qual ha dos reglons de letres grosses de adzur, que dien: "Beatissimo Patri Aleandro Quinto pontifici etc." E mes per lo preu de un drap de pinzell en lo qual esta pintada tota la Ethiopia: los quals libre e drap de manament del Senyor Rey dell son stats comprat en la ciutat de Napols e en lo dit mes de març e foren consignats de continent que foren comprats ço es lo dit drap ala guardaroba del dit senyor en poder de mosser de Mondragon sotscamber del dit Senyor, qui ha carrech special dela dita guardaroba. E lo libre fou consignat ala libreria del dit Senyor en poder dem Thomas Aulesa, guardian dela dita libreria e constaren a preu fet, segons en lo dit albaran se conten que cobre, C. d.*

> *July, 1456. I gave to Tomas Actani, a Florentine merchant who now lives in the city of Naples, 206 ducats and 9 soldi, which were owed to him by virtue of the 2 notarial bills in the following manner, that is, by virtue of the first of the bills, which was written in Naples on the 28th of this past March, for the price of a book called "Cosmografia Tolomei" in large format, written on parchment in ancient script, at the beginning of which there are two big lines of large azure letters that say "Beatissimo Patri Aleandro Quinto pontifici, etc." And another for the price of a cloth on which is painted all Ethiopia. The book and cloth were both, at the request of the Lord King, bought by his estate in the city of Naples in March and were entrusted for keeping after they were purchased, the cloth in the King's wardrobe, which is under the purview of Mr. Mondragon, Chief Steward of the King, who has special care of the wardrobe. And the book was placed in the King's library, which is under the purview of Mr. Thomas Aulesa, who is in charge*

[68] For a biography of Alfonso, a patron of the arts who was quite interested in maps, see Alan F. C. Ryder, *Alfonso the Magnanimous: King of Aragon, Naples, and Sicily, 1396–1458* (Oxford: Clarendon Press; and New York: Oxford University Press, 1990), esp. pp. 314–321 on his library; also see Margaret Ann Skoglund, "Alfonso's Library and his Illuminated Manuscripts," in her "In Search of the Art Commissioned and Collected by Alfonso I of Naples, Notably Painting," Ph.D. Dissertation, University of Missouri - Columbia, 1989, pp. 195–214; Eileen Driscoll, "Alfonso of Aragon as a Patron of Art," in Lucy Freeman Sandler, ed., *Essays in Memory of Karl Lehmann* (Locust Valley, NY: J. J. Augustin, 1964), pp. 87–96; and for discussion of Alfonso's contributions to royal humanism, see Peter Stacey, "Hispania and Royal Humanism in Alfonsine Naples," *Mediterranean Historical Review* 26.1 (2011), pp. 51–65. On the interest of the Aragonese court in geography see Aldo Blessich, *La geografia alla corte aragonese in Napoli: Noticie ed appunti* (Rome: E. Loescher, 1897).

[69] The text quoted here comes from Tammaro de Marinis, *La biblioteca napoletana dei re d'Aragona* (Milan: Hoepli, 1947–1952), vol. 2, pp. 241–242, which came from the *Cedole*, vol. 31, cols. 161b–162b; the translation is my own.

of the library, and they are aware of the price that was paid according to what the bill says he charged C. d.

It had been suggested that the manuscript of Ptolemy in question was BnF MS lat. 4802,[70] that is, one of the three that contains an exemplar of the *Egyptus Novelo* map, as the manuscript was thought to date from the 1450s and did come from the library of the Aragonese kings of Naples, but in fact the manuscript dates from about 1475 to 1480[71] and so cannot be the manuscript whose purchase is described in the passage just quoted. This passage shows that the cloth with the map of Ethiopia was sold in 1456, about 15 years after the Council of Florence, and clearly it existed well before any of the three surviving copies of Ptolemy's *Geography* that contain the *Egyptus Novelo* map (as indicated above, BAV MS Vat. lat. 5699 is dated 1469).

More recently Albinia de la Mare has plausibly suggested that the manuscript that was sold with the map of Ethiopia was London, British Library, Harley MS 7182, which does indeed have the heading at the beginning of the manuscript in two lines of large blue letters, as the entry in the Cedole cited above specifies.[72] De la Mare also suggests that Madrid, Biblioteca Nacional de España, MS Res. 255 was the manuscript of Ptolemy's *Geography* that Vespasiano da Bisticci sent from Florence to Naples to be sold, which is described as *cholla pictura bellissima* and was valued at 50 florins.[73] Harley MS 7182 and BNE MS Res. 255 were produced in the same Florentine workshop and are very similar: the main difference between them is that while the seas on the maps in the London manuscript are painted a rich blue, which could be either azurite or ultramarine (both of which are expensive),[74] the seas in the maps in the Madrid manuscript are painted in a wash that was probably originally green but has faded to a greenish yellow. This pigment, in contrast to the blue of the London manuscript, was inexpensive. But the seas in the Madrid manuscript are deco-

[70]This is suggested by de Marinis, *La biblioteca napoletana dei re d'Aragona* (see the previous note), vol. 2, p. 141, in his description of BnF MS lat. 4802.

[71]On the dating of BnF MS lat. 4802, see the references in note 5 in Chap. 5.

[72]See Albinia Catherine de la Mare, "Observations on Two Italian Manuscripts from Madrid Recently Exhibited in the Bodleian," *Bodleian Library Record* 12.3 (1986), pp. 242–247, at 245.

[73]See de la Mare, "Observations on Two Italian Manuscripts" (see previous note), p. 245; on p. 247 she suggests Escorial MS Vitr. 19 as another possible candidate for the manuscript that Vespasiano da Bisticci sent to Naples to be sold. On the cost of the manuscript in question, see L. Boschetto in his article "Una nuova lettera di Giannozzo Manetti a Vespasiano da Bisticci. Con alcune considerazioni sul commercio librario tra Firenze e Napoli nei decenni centrali del Quattrocento," *Medioevo e Rinascimento* 18 (n.s. 15) (2004), pp. 175–205, at 191. For more on Bisticci see Albina C. de la Mare, "Vespasiano da Bisticci as Producer of Classical Manuscripts in Fifteenth-Century Florence," in Claudine A. Chavannes-Mazel and Margaret M. Smith, eds., *Medieval Manuscripts of the Latin Classics: Production and Use* (Los Altos Hills, CA: Anderson-Lovelace, 1996), pp. 167–208, esp. 201 on the sale of the Ptolemy in Naples.

[74]On the prices of azurite and ultramarine, see Susanne Kubersky-Piredda, "The Market for Painters' Materials in Renaissance Florence," in Jo Kirby, Susan Nash, and Joanna Cannon, eds., *Trade in Artists' Materials: Markets and Commerce in Europe to 1700* (London: Archetype, 2010), pp. 223–243, esp. 225 and 238–239.

rated with a remarkable number and variety of sea monsters,[75] which must have added considerably to the cost of the manuscript. As the two manuscripts are of similar sizes, and the one had expensive paint used for its seas but no sea monsters, while the other has inexpensive paint used for its seas but expensive sea monsters, it seems reasonable to assume that the original prices of the two manuscripts were not terribly far apart.

The Madrid manuscript was valued at 50 Florentine florins, while the London manuscript together with the map of Ethiopia cost just over 206 Neapolitan ducats. The Florentine *fiorino a oro* was nominally pure gold, while the ducat of Naples, which was equivalent in weight to the Florentine florin, in the fifteenth century was 97.5% pure,[76] so the 206 Neapolitan ducats were worth just over 200 Florentine gold florins.[77] We have additional data regarding the cost of a mid-fifteenth-century Florentine manuscript of Ptolemy's *Geography*: there are records of payments from Álvaro Alfonso, the Bishop of Algarve, to Pietro del Massaio and Giovanni de Maestro Antonio (probably a copyist), *per parte d'uno libro delle tavole di Tolomeo*, "for the part of a book with the maps of Ptolemy," and the payments totaled 63 florins and 33 soldi.[78] These payments seem to have been just for the maps (possibly with brief descriptions from Book 8 of the *Geography* on the backs of the maps, as is typical of Pietro del Massaio's manuscripts of this period), rather than for a complete manuscript of the *Geography*, i.e., complete with the text of Books 1 to 7. The higher price compared to those of the two other manuscripts just mentioned—particularly given that the price was just for the maps, rather than for a complete manuscript of the *Geography*—may reflect a larger map size,[79] additional decorative elements such as images of cities or gold leaf, or both.

[75]See Chet Van Duzer, "The Sea Monsters in the Madrid Manuscript of Ptolemy's *Geography* (Biblioteca Nacional, MS Res. 255)," *Word & Image* 27.1 (2011), pp. 115–123; and Chet Van Duzer, *Sea Monsters on Medieval and Renaissance Maps* (London: British Library, 2013), pp. 61–66.

[76]On the purity of the Neapolitan ducat in the fifteenth century, see Philip Grierson and Mark Blackburn, *Medieval European Coinage: with a Catalogue of the Coins in the Fitzwilliam Museum* (Cambridge, UK, and New York: Cambridge University Press, 1986–1998), vol. 14, p. 451.

[77]On the value of the ducat of Naples also, see Peter Spufford, *Handbook of Medieval Exchange* (London: Offices of the Royal Historical Society, 1986), p. 64.

[78]The records of these payments are in Florence, Archivio dello Spedale degli Innocenti, Fondo Estranei 248, f. 162r, and are discussed by Gentile, *Firenze e la scoperta dell'America* (see note 8 in Chap. 1), pp. 200–202 (who, however, adds the sums incorrectly, arriving at 53 florins and 33 soldi in place of 63 florins); and Angelo Cattaneo, *Fra Mauro's Mappa Mundi and Fifteenth-Century Venice* (Turnhout: Brepols, 2011), pp. 286–287.

[79]Harley MS 7182 and BNE MS Res. 255 are large, 58.5 × 46 cm and 56 × 41 cm, respectively, but there are substantially larger manuscripts of Ptolemy's *Geography*, including BnF MS lat. 4802, 88 × 59.5 cm; Salamanca, Biblioteca General de la Universidad, MS 2586, 88 × 57 cm; BAV Vat. Lat. 5699, 76 × 49.5 cm; and BAV Urb. lat. 277, 70.5 × 43.5 cm. For an engaging discussion of large medieval manuscripts, see George D. Greenia, "The Bigger the Book: On Oversize Medieval Manuscripts," *Revue belge de philologie et d'histoire* 83.3 (2005), pp. 723–745.

We also have records that show that on March 30, 1466, Borso d'Este, Duke of Ferrara (1413–1471),[80] paid 100 ducats directly to the cartographer Donnus Nicolaus Germanus[81] for a manuscript of Ptolemy's *Geography*.[82] The manuscript survives and is in Modena, Biblioteca Estense Universitaria, MS Lat. 463 = α.X.1.3; it measures 446 × 305 mm and has been reproduced in facsimile.[83] Given that the Modena manuscript is substantially smaller than BNE MS Res. 255 (56 × 41 cm), it is remarkable that the Modena manuscript sold for twice the price of the Madrid manuscript. The most likely explanation is that the seas on the maps in the Modena manuscript are painted in ultramarine rather than azure, though it is also possible that Borso d'Este or his agents simply did not negotiate the price of the manuscript well.[84]

There is also a record that Alfonso V of Aragon, King of Naples, purchased a manuscript of Ptolemy's *Geography* in 1453, while he was in Fontana, near Pozzuoli, for the price of 170 ducats, 11 tarì, and 10 grani,[85] the equivalent of about 166 Florentine florins. This manuscript was extraordinarily expensive, and this price must reflect a very large map size and an extravagant program of decoration, with (e.g.) seas painted in

[80]On Borso d'Este's activities as a patron of the arts, see Lubomyra Adrianna Lesychyn, "The Magnificence of Borso d'Este, 1450–1471," in her "The Magnificence of Borso and Ercole d'Este Princes of Ferrara, 1450–1505," M.A. Thesis, McMaster University (Canada), 1981, pp. 33–69; Luke Syson, "Lo stile di una signoria: il mecenatismo di Borso d'Este," in Mauro Natale, ed., *Cosmè Tura e Francesco del Cossa: l'arte a Ferrara nell'età di Borso d'Este* (Ferrara: Ferrara arte, 2007), pp. 75–88; and Ulrike Bauer-Eberhardt, "Zur ferraresischen Buchmalerei unter Borso d'Este: Taddeo Crivelli, Giorgio d'Alemagna, Leonardo Bellini und Franco dei Russi," *Pantheon* 55 (1997), pp. 32–45.

[81]For references on Nicolaus Germanus, see note 5 in Chap. 1 above.

[82]On Borso d'Este's payment of 100 ducats for the manuscript of Ptolemy's *Geography*, see Herman Julius Hermann, "Zur Geschichte der Miniaturenmalerei am Hofe d'Este in Ferrara," *Jahrbuch der kunsthistorischen Sammlungen des allerhöchsten Kaiserhauses* 21 (1900), pp. 114–271, at 261; cited in Józef Babicz, "The Celestial and Terrestrial Globes of the Vatican Library, Dating from 1477, and their Maker Donnus Nicolaus Germanus (ca 1420–ca 1490)," *Der Globusfreund* 35–37 (1987), pp. 155–168, at 161.

[83]The facsimile is Ptolemy, *Cosmographia* (Modena: Il Bulino, 2004); it is accompanied by a volume of commentary by Laura Federzoni and Annalisa Battini, *L'atlante di Borso d'Este: la Cosmographia di Claudio Tolomeo della Biblioteca Estense Universitaria di Modena: commentario all'edizione in facsimile del codice miniato a.X.1.3 = Lat. 463* (Modena: Il Bulino, 2006). The commentary volume, which is more widely distributed than the facsimile itself, includes reduced-scale reproductions of all of the folios of the manuscript. There is also discussion of the manuscript in Joseph Fischer, *Claudii Ptolemaei Geographiae, Codex Vrbinas Graecvs 82* (Leiden: Brill, and Leipzig: Harrassowitz, 1932), vol. 1.1, pp. 215 and 344–347 and Cavallo, *Cristoforo Colombo e l'apertura degli spazi* (see note 42 in Chap. 1), vol. 1, pp. 207 and 212, with an illustration on pp. 210–211.

[84]On the question of royals or nobles not negotiating the price of a map well, in July of 1511, Marquis Francesco II Gonzaga's offered of 50 ducats—sight unseen—for maps of Cairo and Jerusalem to their owner in Venice. The offer was not accepted by the owner of the maps, and when the Marquis's agents Carlo Valier and one Maestro Lorenzo saw the maps, they judged that they were not worth more than a single ducat. On this episode see Clifford M. Brown, "The 'Camera del Mapamondo et del Caiero' in the Palazzo di San Sebastiano in Mantua: A Fragment of a View of Jerusalem and Vittorio Carpaccio's Letter to Francesco II Gonzaga of 1511," *Journal of Jewish Art* 10 (1984), pp. 32–46, at 39.

[85]See de Marinis, *La biblioteca napoletana dei re d'Aragona* (see note 69 in Chap. 5), vol. 2, pp. 141 and 237, citing the Cedole della Tesoreria Aragonese, vol. 24, f. 346r.

ultramarine,[86] images of cities, perhaps some *tabulae modernae*, and extensive application of gold leaf.

Assuming it was Harley 7182 that was sold with the map of Ethiopia, and given that Harley 7182 would seem to have been of more or less the same value as BNE MS Res. 255, i.e., 50 florins, that would mean that the map of Ethiopia was valued at about 150 florins—a staggering sum. By way of comparison, Afonso V, King of Portugal, in 1457, paid 28 Venetian ducats (which had precisely the same value as Florentine florins) to commission a copy of Fra Mauro's world map, which is extremely detailed and colorful and measures about two meters in diameter.[87] Also, according to contracts signed in Barcelona in 1399, the cartographers Jacme Ribes and Francesco Becaria were to receive 320 florins of Aragon for producing 4 world maps, 120 for 2 smaller *mappaemundi*, each of which was to be about 310 × 155 cm, and 200 for 2 larger *mappaemundi*, each of which was to be about 368 × 368 centimeters. All of the maps were richly decorated, and the latter two were each larger than the Ebstorf *mappamundi*, which until its destruction in World War II was the largest surviving medieval map. The florin of Aragon was worth about 75% of the Florentine florin, so Jacme Ribes and Francesco Becaria were to be paid the equivalent of 75 Florentine florins for each of the large world maps.[88] This is much less than the 150 florins (equivalent) apparently paid for the map of Ethiopia.

But returning to the purchase of the manuscript of Ptolemy and the map of Ethiopia, let us assume for the sake of argument that the manuscript in question was not Harley MS 7182, and not similar to BNE MS Res. 255, which was valued at 50 florins; let us assume instead that it was a larger and more elaborately decorated manuscript. Even if it were worth 100 florins—twice as much as BNE MS Res. 255, and an amount which is a reasonable estimate of the cost of a hypothetical complete manuscript of

[86] The blue paints made from azurite and ultramarine are visually indistinguishable, but while azurite was expensive, ultramarine typically cost ten times as much. If ultramarine was to be used in painting a manuscript, this was often specified in the contract, and the commissioner of the manuscript often made a separate payment for the ultramarine. Thus the painting of the seas with ultramarine in the twenty-seven standard maps in a manuscript of Ptolemy would very substantially increase the cost of the manuscript. On the relative costs of azurite and ultramarine, see Susanne Kubersky-Piredda, "The Market for Painters' Materials in Renaissance Florence," in Jo Kirby, Susan Nash, and Joanna Cannon, eds., *Trade in Artists' Materials: Markets and Commerce in Europe to 1700* (London: Archetype, 2010), pp. 223–243, esp. 225 and 238–239; and Andreas Burmeister and Christoph Krekel, "The Relationship between Albrecht Dürer's Palette and Fifteenth/Sixteenth Century Pharmacy Price Lists: The Use of Azurite and Ultramarine," in Ashok Roy and Perry Smith, eds., *Contributions to the Dublin Congress, 7–11 September 1998: Painting Techniques: History, Materials and Studio Practice* (London: International Institute for Conservation of Historic and Artistic Works, 1998), pp. 101–105, esp. 102–104.

[87] For full discussion and bibliography, see Angelo Cattaneo, *Fra Mauro's Mappa Mundi* (see note 78 in Chap. 5), pp. 282–284.

[88] The contracts with Jacme Ribes and Francesco Becaria are discussed by R. A. Skelton, "A Contract for World Maps in Barcelona, 1399–1400," *Imago Mundi* 22 (1968), pp. 107–113; my discussion here is based on that of Angelo Cattaneo, *Fra Mauro's Mappa Mundi* (see note 78 in Chap. 5), pp. 292–296, who supplies bibliography regarding the relative values of the florin of Aragon and the Florentine florin.

Ptolemy's *Geography* that included the maps for which Álvaro Alfonso paid 63 florins and 33 soldi, and exactly the price that Borso d'Este paid for his manuscript of the *Geography*—the map of Ethiopia would still in effect be valued at 100 florins, much more than the contract price of 75 florins for the very large and richly decorated *mappamundi* to be made by Jacme Ribes and Francesco Becaria, and close to 4 times as much as a copy of Fra Mauro's large and elaborate world map. It is unlikely that the manuscript of Ptolemy that Alfonso V of Aragon bought in 1456 had the same value as the one he bought in 1453, namely, 170 ducats, 11 tarì, and 10 grani, in part because the manuscript he bought in 1456 would seem to be Harley 7182 but also because such expensive manuscripts of Ptolemy must have been rare. But even if it did have that high value, the map of Ethiopia would still have a value of almost 30 ducats—more than the copy of Fra Mauro's splendid *mappamundi*.

The map of Ethiopia sold in 1456, then, was an extraordinary object, possibly the most expensive map of the fifteenth century but in any case almost certainly valued more highly than one of the most detailed and artistically elaborate world maps of its era (Fra Mauro's). This map cannot merely have been a version of a Ptolemaic map of Ethiopia. In his *Geography*, Ptolemy shows little in *Ethiopia sub Aegypto* beside the southern course of the Nile, and in *Aethiopia Interior* there is nothing beside the Mountains of the Moon, from which the Nile arises and five mountains and one river in the western part of the continent (Fig. 5.6). To the best of our knowledge, there were only two sets of cartographic information about Ethiopia that were available in Europe and detailed enough to generate a truly impressive map of Ethiopia: the archetype of the *Egyptus Novelo* maps and the Ethiopian maps used by Fra Mauro. As indicated above, there were other interactions between Europe and Ethiopia in the fifteenth century through which detailed geographical and cartographical information could have been transmitted to Europe,[89] but the most likely possibility is that the map of Ethiopia sold in 1456 was the original and more complete version of the *Egyptus Novelo* map, the version that extended its depiction east to include the southeastern part of Africa to the ocean and that in making his Yale map, Martellus made use either of that map or one derived from it.

It should be noted that the purchase of the map of Ethiopia by Alfonso V of Aragon was not merely a royal whim, for Alfonso had a strong interest in expanding in the East and in forging an alliance in Ethiopia.[90] In 1427 or 1428, as King of Valencia (1416–1458), he received an embassy from Ethiopia, and Alfonso sent an embassy in return with instructions to

[89]On early modern contacts and interactions between Ethiopia and Europe see the references in note 59 in Chap. 5.

[90]Francesco Cerone, "La politica orientale di Alfonso di Aragona," *Archivio storico per le provincie napolitane* 27 (1902), pp. 1–93, 380–456, 555–634, and 774–852; and 28 (1903), pp. 154–212.

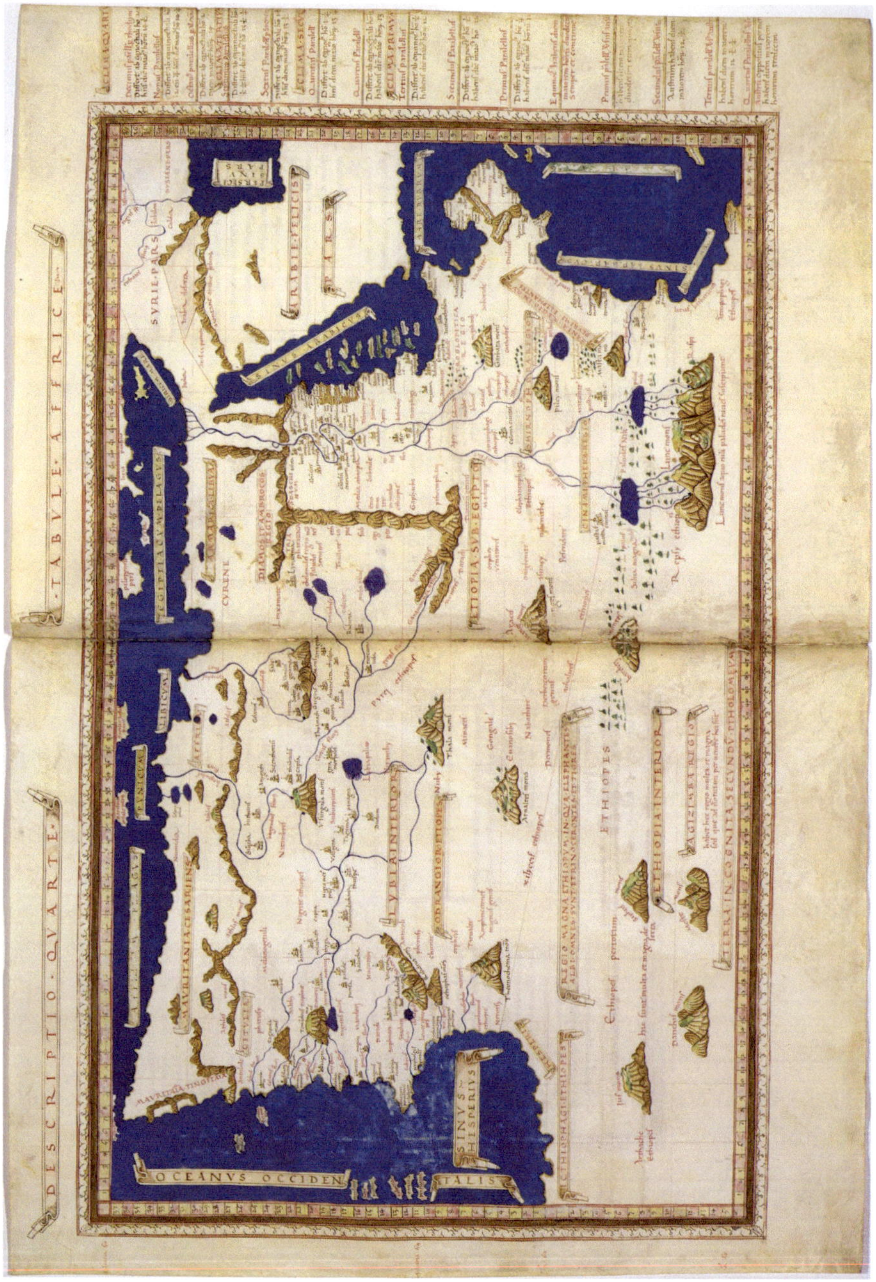

Fig. 5.6 Ptolemy's fourth map of Africa, showing the lack of detail in the southern part of the continent, in a manuscript of Ptolemy's Geography *whose maps were made by Martellus (Florence, BNCF, Magliabechiano XIII 16, ff. 130v–131r). By concession of the Ministero dei Beni e delle Attività Culturali e del Turismo/Biblioteca Nazionale Centrale, Firenze. (Further reproduction in any medium is forbidden)*

explore the possibility of cementing an alliance through a royal marriage—though the embassy never reached its destination.[91] Some years later he was in diplomatic contact with Zar'a Yâ'eqōb, the Christian sovereign of Ethiopia (1434–1468), whose interests corresponded well with those of European Christians; for example, in 1443 Zar'a Yâ'eqōb had sent an embassy to the Mamluk Sultan in Cairo to warn him not to persecute the

[91] See Peter P. Garretson, "A Note on Relations between Ethiopia and the Kingdom of Aragon in the Fifteenth Century," *Rassegna di studi etiopici* 37 (1993), pp. 37–44.

Copts. In 1450 he sent an embassy to Alfonso to propose a joint crusade to retake the Holy Land, entrusting this mission to Pietro Rombolo, who was mentioned earlier. Alfonso promised to send 150 warships for the conquest of the Holy Land when he received word that Zar'a had moved his army to the southern border of Egypt and had diverted the course of the Nile.[92] Alfonso sent additional embassies to Zar'a Yâ'eqōb in 1452 and 1453 but of a more perfunctory nature[93]; nonetheless, his purchase of the map of Ethiopia in 1456 certainly grew out of his long-standing interest in the region, and this interest explains his willingness to pay a very high price for the map.

Unfortunately there does not seem to be any evidence regarding how the information on the map of Ethiopia purchased by Alfonso in 1456 and taken to his palace in Naples might have been transmitted to Martellus in Florence in the late 1480s or early 1490s: whether the map was copied, whether Martellus might have traveled to Naples, or there was some other method of transmission.

Our study of southern Africa on the Yale Martellus map has emphasized the uniqueness and great value of that map: it is evidently the only surviving copy of a more complete version of the *Egyptus Novelo* map, a document of tremendous importance in the history of indigenous mapping of Africa, which was also important in advancing European knowledge of the continent. In addition, the map is an intriguing addition, albeit in an attenuated way, to the corpus of maps that locate the Terrestrial Paradise in southern Africa.

[92] I am paraphrasing here from Eliyahu Ashtor, "Alfonso il Magnanimo e i Mamlucchi," *Archivio Storico Italiano* 142 (1984), pp. 3–39, at 24–25; also see Cerone, "La politica orientale di Alfonso di Aragona" (see note 90 in Chap. 5), pp. 39–40 and 64–65.

[93] Cerone, "La politica orientale di Alfonso di Aragona" (see note 90 in Chap. 5), pp. 76–80.

Chapter 6

The Influence of the Yale Martellus Map

In the preceding pages, I have brought forward previously unnoticed evidence regarding the influence of the Yale Martellus map on Giovanni Contarini's world map of 1506 and have demonstrated the profound influence of the map—or one very similar to it—on Martin Waldseemüller's world map of 1507. In the past, it has been suggested that the Yale Martellus map heavily influenced Martin Behaim in the creation of his terrestrial globe of 1492,[1] but this claim has not been examined in detail, and I propose to do so now.

There are several elements of the Yale Martellus map that are shared by Behaim's globe, which make it essentially certain that Behaim had either the Yale Martellus map or another very similar to it that he used as a source in creating his globe. I will focus on elements that are shared by the Yale Martellus map and Behaim's globe, but do not appear on Martellus's other surviving world maps: these elements point specifically to a large world map like the Yale Martellus map. All of Martellus's world maps have the distinctive so-called fourth peninsula or Dragon's Tail peninsula in eastern Asia, but the Yale Martellus map and Behaim's globe share a well-defined corner in that peninsula just south of the equator: as one follows the eastern coast of the peninsula to the south, it runs almost straight to the south, and then just past the equator, it takes a sharp turn to the southwest (see Figs. 1 and 6.1). Another feature in eastern Asia shared by the Yale Martellus map and Behaim's globe, but not by any of Martellus's other world maps, is a triangular peninsula on the eastern coast of Asia that juts eastward into what we would call the Pacific. On both the Yale Martellus map and on Behaim's globe, the southern edge of this peninsula runs almost exactly west and east and is just north of the Tropic of Cancer (Figs. 1 and 6.2). On both maps, Japan is oriented to the north and south, and the Tropic of Cancer strikes it about one third of the way down from its northern tip.

Five islands in the southern Indian Ocean on Behaim's globe—Seilan, Pentan, Neueuocan, Java, and Java Minor—appear on both the Yale Martellus map and on the Martellus-Rosselli map, but on the latter, they are located much closer to the so-called fourth peninsula than they are on the Yale map or Behaim's globe. These locations on the Martellus-Rosselli map are perhaps the result of fitting the islands in the map's less ample confines, but in any case, the Yale map or one very similar to it is a much more likely source for the positions of these islands relative to the mainland than the Martellus-Rosselli map. In addition, Behaim has the island of Candyn east of Java Major, and the Yale Martellus map has an island in the same position (though no name is legible on it), while no such island appears on any of Martellus's other surviving maps. Further, on Behaim's globe, there are two substantial unnamed islands just south of the Tropic of Capricorn between Madagascar and the tip of the fourth peninsula, and

[1] For earlier discussions of the influence of the Yale Martellus map on Behaim's globe, see note 3 in Front Matter above.

C. Van Duzer, *Henricus Martellus's World Map at Yale (c. 1491)*, Historical Geography and Geosciences,
https://doi.org/10.1007/978-3-319-76840-3_6

Fig. 6.1 The Fourth Peninsula and southeast Asian islands on Behaim's globe of 1492 (Nürnberg, Germanisches Nationalmuseum, WI 1826). (By permission of the Germanisches Nationalmuseum)

there are two substantial islands (surrounded by several smaller ones) in similar positions on the Yale Martellus map,[2] but only one on the Martellus-Rosselli map, so again, the Yale Martellus map or one very similar to it seems like the more likely source.

Another intriguing link between the Yale Martellus map and Behaim's globe is that both have some sea monsters that are derived from the illustrated encyclopedia titled *Hortus Sanitatis*, which was first published in 1491.[3] Earlier I discussed Martellus's legends about sea monsters that are derived from the *Hortus Sanitatis*, and Behaim has illustrations of three sea monsters that seem to be derived from that book. The pose of the two siren-like creatures south of the Cape Verde Islands on Behaim's globe (Fig. 6.3) is very similar indeed to that of the two siren-shaped dolphins in the *Hortus Sanitatis*, "De piscibus," Chapter 27 (Fig. 6.4)—so much so that it is impossible to doubt that Behaim was using either the *Hortus Sanitatis* or a work dependent on that book. East of the sirens, there was a rampant aquatic lion depicted simply as a lion in the sea which is now faded almost

[2]The two islands between Madagascar and the fourth peninsula on the Yale Martellus map do not seem to have names, but it is quite possible that the names have simply faded to illegibility. On Waldseemüller's 1507 map, the two corresponding islands are labeled *Iona* and *Callenzuam*.

[3]For references on the *Hortus Sanitatis*, see note 100 in Chap. 1.

Fig. 6.2 The triangular peninsula on the eastern coast of Asia on Behaim's globe of 1492 (Nürnberg, Germanisches Nationalmuseum, WI 1826). (By permission of the Germanisches Nationalmuseum)

to invisibility (Fig. 6.5), but its pose is similar to that of the aquatic lion in the *Hortus Sanitatis*, "De piscibus," Chapter 48, which is depicted as a lion with scales (Fig. 6.6).[4] The use of the *Hortus Sanitatis* as a source by both Martellus in his Yale map and Behaim in his globe is yet another link between these two works. There are no details that are derived from the *Hortus Sanitatis* in any of Martellus's other surviving world maps.

At the same time, there are various details on Behaim's globe that indicate that he was using sources other than the Yale Martellus map. For instance, the systems of mountain ranges in the interior of Africa are dif-

[4] I discuss Behaim's use of the *Hortus Sanitatis* as an iconographical source for his dolphins and sea lion in my *Sea Monsters on Medieval and Renaissance Maps* (London: British Library, 2013), pp. 69–70.

Fig. 6.3 Two siren-like creatures in the Atlantic on Behaim's globe (Nürnberg, Germanisches Nationalmuseum, WI 1826). (By permission of the Germanisches Nationalmuseum)

Fig. 6.4 Dolphins from the 1491 edition of the Hortus santitatis, *"De piscibus," chapter 27. Library of Congress, Rare Book and Special Collection Division, Incun. 1491 .H75. (Courtesy of the Library of Congress)*

Fig. 6.5 The aquatic lion in the Atlantic on Behaim's globe, from E. G. Ravenstein, Martin Behaim, His Life and His Globe *(London: G. Philip & Son, Ltd., 1908). (Courtesy of the David Rumsey Historical Map Collection)*

Fig. 6.6 The aquatic lion in the 1491 edition of the Hortus sanitatis, *"De piscibus," chapter 48. Library of Congress, Rare Book and Special Collection Division, Incun. 1491 .H75. (Courtesy of the Library of Congress)*

ferent on the map and the globe; as detailed earlier, many of the place names on the western coast of Africa differ between the map and the globe; the shape of the island of Taprobana is different; Behaim has an island named Marisa on the equator in the Indian Ocean east of Africa, and a substantial gulf in the African coast just to the west of it, neither of which appears on the Yale Martellus map; and he has a number of place names in Arabia that do not come from Ptolemy[5] and thus do not appear on the Yale

[5] For the non-Ptolemaic toponyms in Arabia on Behaim's globe, see Ravenstein, *Martin Behaim, His Life and His Globe* (see note 72 in Chap. 1), p. 82.

Martellus map. It is not at all surprising that Behaim should use sources other than the Yale Martellus map in creating his globe, but some of the details on his globe point to Martellus, albeit not to the Yale map, making it seem likely that Behaim was using a different version of Martellus's large map than the one at Yale.

For example, on Behaim's globe in southern Africa, there is a peninsula jutting eastward into the Indian Ocean, but its shape is not particularly similar to the shape of the eastward jutting peninsula on the Yale Martellus map. In fact, the shape of southern Africa on Behaim's globe (Fig. 6.7) is strikingly similar to that in the world map in the Florence manuscript of Martellus's *Insularium* (see Fig. 1.2): in both cases, if we follow the coastline from west to east, it cuts down to the southeast, then proceeds directly east for some distance, and then curves to the northeast to form the southern coast of the eastward jutting peninsula. Also, on Behaim's globe, the island of Madagascar has a distinctive "L" shape that it lacks on the Yale Martellus map and is also further away from the African mainland than it is on the Yale Martellus map. Yet the island does have this "L" shape on the Martellus-Rosselli map (see Fig. 1.6)—and it has the same shape on Waldseemüller's 1507 map.[6] Similarly, the island of Zanzibar appears to the south of Madagascar on Behaim's globe; it does not appear at all on the Yale map, but it does appear south of Madagascar on the Martellus-Rosselli map, and also on Waldseemüller's 1507 map.[7] Also, while the disposition of the islands Seilan, Pentan, Neueuocan, and Java on Behaim's globe is generally similar to that on the Martellus-Rosselli map, there are important differences between the ways the islands are placed on the Yale Martellus map and Behaim's globe, so Behaim was evidently not using the Yale map but rather another large and detailed map from Martellus's hands.

Thus, while Behaim's globe clearly shows the influence of Martellus's maps, and he certainly had access to a large, elaborate map very similar to the Yale Martellus map, we can be almost certain that the map that he was consulting was different from the Yale map: this is a further evidence that Martellus produced more than one large world map and that the version that Behaim was using extended somewhat further to the south than the

[6]There is another point of close similarity between Behaim's globe and Waldseemüller's 1507 map that should be taken as indicating the influence of Martellus: Ravenstein, *Martin Behaim*, p. 93, restores a damaged legend on the globe thus, *hie f[indt man rohr] von x [spanne di rund u 15] schritt*, "Here are found canes ten spans round and fifteen paces long," based on a legend in the same spot on Waldseemüller's map, which reads *hic sunt canne grosse tres spanne et longe 0.15. passus*. The appearance of what is the same legend in the same spot on Behaim's globe and Waldseemüller's map is due to the influence of Martellus: as indicated above, there is a very similar legend in the same location on the Yale Martellus map that reads in part *canne grosse tress spanne et longes quindecim passus*. Incidentally, Madagascar also has the "L" shape in the world maps in *Der Welt Kugel* (Strasbourg: Grüninger, 1509) and *Globus mundi* (Strasbourg: Grüninger, 1509), which have been attributed to Waldseemüller.

[7]Marco Polo does not clearly indicate the position of Zanzibar relative to Madagascar, saying only that the former is "beyond" the latter, so the identical locations of Zanzibar on Behaim's globe, the Martellus-Rosselli map, and Waldseemüller's 1507 map cannot be the product of coincidence: they are evidence of Martellus's influence.

Fig. 6.7 *Southeastern Africa on Mcrtin Behaim's globe of 1492 (Nürnberg, Germanisches Nationalmuseum, WI 1826). (By permission of the Germanisches Nationalmuseum)*

Yale map; it gave southern Africa a shape like that in the world map in the manuscript of Martellus's *Insularium* now in Florence (and so did not show the influence of the *Egyptus Novelo* map), and perhaps had illustrations of sea monsters based on the *Hortus Sanitatis* (rather than just the *Hortus Sanitatis*-derived texts about sea monsters that are on the Yale map).

There is another intriguing similarity between the Yale Martellus map and Behaim's globe: their extensive use of Marco Polo.[8] Around the middle of the fifteenth century, Fra Mauro made heavy use of Marco Polo in the creation of his *mappamundi*,[9] and geographers and cartographers would continue to rely on Polo for more than a century after Fra Mauro.[10] It is entirely possible that Behaim and Martellus came independently to their heavy use of Polo in eastern Asia, but given Behaim's following of Martellus in the outlines of eastern Asia, it is tempting to think that he was also inspired by Martellus's example to take place names and legends in this region from Polo. Behaim's legends on the islands in the Indian Ocean and also on Japan have been transcribed and translated,[11] and several of the legends in this area on the Yale Martellus map were transcribed above, as were the legends on Japan on the map of that island in the Florence manuscript of Martellus's *Insularium*.[12] In these cases, unless Martellus produced a large world map that was quite different from the Yale Martellus map in terms of his use of Marco Polo, Behaim seems not to have been following Martellus closely. Behaim's legends on Japan are different from Martellus's on the Yale map, and Behaim's legends about islands in the Indian Ocean are substantially longer than Martellus's. Also, Behaim several times cites a specific book and chapter in Polo from which he takes his information,[13] and this style of citation does not appear on the Yale

[8] On Behaim's use of Marco Polo, see Ravenstein, *Martin Behaim, His Life and His Globe* (see note 72 in Chap. 1), p. 63.

[9] On Fra Mauro's use of Marco Polo, see G. R. Crone, "Fra Mauro's Representation of the Indian Ocean and the Eastern Islands," *Studi Colombiani* 3, 1952, pp. 57–64; Angelo Cattaneo, "Fra Mauro *Cosmographus incomparabilis* and his *Mappamundi*: Documents, Sources, and Protocols for Mapping," in Diogo Ramada Curto, Angelo Cattaneo, and André Ferrand Almeida, eds., *La cartografia europea tra primo Rinascimento e fine dell'Illuminismo: Atti del convegno internazionale, The Making of European Cartography (Firenze, BNCF-EUI, 13–15 dicembre 2001)* (Florence: L. S. Olschki, 2003), pp. 19–48, esp. 39; Falchetta, *Fra Mauro's World Map* (see note 85 in Chap. 1), pp. 61–67; Eugenio Burgio, "'Cartografie' del viaggio: Sulle relazioni fra la 'Mappamundi' di Fra Mauro e il 'Milione'," *Critica del testo* 12.1 (2009), pp. 59–106; and Angelo Cattaneo, *Fra Mauro's Mappa Mundi* (see note 78 in Chap. 5), pp. 191–198 and 211–219.

[10] On the reception of Polo's work, see Christine Gadrat-Ouerfelli, *Lire Marco Polo au Moyen Age: traduction, diffusion et réception du Devisement du monde* (Turnhout: Brepols, 2015).

[11] See Ravenstein, *Martin Behaim, His Life and His Globe* (see note 72 in Chap. 1), pp. 86–89 and 104–105.

[12] For the similarity between the legends on Japan on the Yale Martellus map and on the map of Japan in the Florence manuscript of Martellus's island book, see above pp. XX–YY.

[13] For examples of Behaim's legends in which he cites specific passages in Marco Polo, see Ravenstein, *Martin Behaim, His Life and His Globe* (see note 72 in Chap. 1), p. 86 on Seilan, p. 87 on Java Minor and Pentan, p. 88 on Nekuran and Angaman, p. 89 on Japan, p. 104 on Zanzibar, and p. 105 on Madagascar and Scotra. On Behaim's use of Polo see Gadrat-Ouerfelli, *Lire Marco Polo au Moyen Age* (see note 10 in Chapt. 6), pp. 315-318.

Martellus map. Thus it seems quite possible that Behaim drew inspiration from Martellus's use of Marco Polo, so that we may tentatively count this as another case of Martellus's influence on Behaim, but it does not seem that Behaim was simply copying Martellus's Polo-derived legends.

Then there is the question of the influence of the Yale Martellus map (or one very similar to it) on Columbus. There is good evidence that Martellus map either inspired or reflects Columbus's geographical ideas. In the following paragraphs, I will discuss the evidence of congruence between Columbus's geographical ideas and the geography of the Yale Martellus map briefly brought forward by Skelton in his unpublished paper on the map, adding considerably more detail and also taking issue with some points.[14]

First, Skelton notes that Columbus's son Fernando in Chapter 20 of his biography of his father said that Columbus certainly would have found Cipangu (Japan) if he had not believed the report that the island extended from north to south[15]: according to Fernando, as a result of believing this report, Columbus did not sail far enough south to find Cipangu.[16] I know of no early text that says that Japan stretched from north to south—in particular, Marco Polo, who was the only available textual source on Japan in the late fifteenth century, makes no such claim.[17] Moreover, it does not have that orientation on Fra Mauro's *mappamundi* of c. 1455, which is one of the few maps earlier than Martellus's to depict the island.[18] But that is precisely how the island is depicted in the Yale Martellus map—that is, oriented to the north and south (the map of Japan in the Florence manuscript of Martellus's *Insularium*—see Fig. 1.9—lacks any indication of orientation).

Second,[19] on his Fourth Voyage in 1502–1504, as Columbus was sailing down the coast of Central America, he identified what is now Honduras as Ciamba in eastern Asia.[20] He was told about a wealthy region called Ciguare, and he writes: "They also say that the sea surrounds Ciguare, and

[14]R. A. Skelton, "World Map by Henricus Martellus Germanus, c. 1489, at Berne," January 10–17, 1960 (see note 2 in Front Matter), pp. 14–17. For references to other discussions of the possible links between the Yale Martellus map and Columbus, see note 4 in Front Matter.

[15]Skelton, "World Map by Henricus Martellus Germanus, c. 1489, at Berne" (see note 2 in Front Matter), p. 16.

[16]Fernando Colón, *The History of the Life and Deeds of the Admiral Christopher Columbus: Attributed to His Son Fernando Colón*, ed. Ilaria Luzzana Caraci, trans. Geoffrey Symcox and Blair Sullivan (Turnhout: Brepols, 2004), pp. 69–70 (English) and 271 (Italian).

[17]For Marco Polo's description of Japan, see Marco Polo, *Marka Pavlova z Benátek, Milion* (see note 57 in Chap. 2), Book 3, Chapter 2, pp. 153–154, and Yule, *The Book of Ser Marco Polo* (see note 44 in Chap. 2), Book 3, Chapter 2, vol. 2, pp. 253–257.

[18]George Kish supplies the outline of Japan on Fra Mauro's map in "Two Fifteenth-Century Maps of 'Zipangu': Notes on the Early Cartography of Japan," *The Yale University Library Gazette* 40.4 (1966), pp. 206–214, on p. 211.

[19]Skelton, "World Map by Henricus Martellus Germanus, c. 1489, at Berne" (see note 2 in Front Matter), p. 16, but I correct and expand substantially on Skelton's hints.

[20]See Christopher Columbus, *Select Letters of Christopher Columbus, With Other Original Documents, Relating to His Four Voyages to the New World*, trans. R. H. Major (London: Hakluyt Society, 1870), pp. 180.

that at ten days' journey from thence is the river Ganges. These lands appear to have the same bearings with respect to Veragua as Tortosa has to Fontarabia, or Pisa to Venice."[21] By this statement he seems to mean that Ciguare and the river Ganges are on the sea on the far side of the land that he is coasting down.[22]

These beliefs are consistent with the idea that as he was sailing south along the coast of Central America, Columbus imagined himself sailing down the huge peninsula in eastern Asia on Martellus map, thinking that he could reach the Ganges after he rounded its southern tip. In about 1506, the Venetian cosmographer Alessandro Zorzi made three sketch maps that represent the geographical views of the Columbus brothers.[23] In one of the maps, Zorzi places New World toponyms from Columbus's Fourth Voyage down the eastern coast of a large peninsula in Asia (Fig. 6.8).[24] This map certainly seems to show the influence of the Yale Martellus map or one very similar to it, for the large peninsula that juts eastward from Asia is very similar indeed to the large eastward-jutting peninsula on Martellus map.

Third, Skelton refers[25] to a passage from the *pleitos colombinos* or Columbian lawsuits of August 11, 1515, in which the son of Martín Alonso Pinzón (the father had sailed with Columbus on his First Voyage) records that his father, in or before 1492, saw a document in the Vatican Library that indicated that Cipangu lay 95° west of Spain by an easy

[21] See Columbus, *Select Letters* (see previous note), pp. 182.

[22] Skelton, "World Map by Henricus Martellus Germanus, c. 1489, at Berne" (see note 2 in Front Matter), p. 16, suggests that Columbus thought that he was heading for Catigara, a city in Southern Asia. Columbus does mention Catigara in his account of his Fourth Voyage, but it is not at all clear that this was a destination or even a waypoint that he had in mind: see Columbus, *Select Letters* (see note 20 in Chap. 6), p. 183.

[23] The maps are Florence, Biblioteca Nazionale Centrale, Banco Rari 234, ff. 56v, 57r, and 60v. In the early literature, Zorzi's three sketch maps were attributed to Bartolomew Columbus: see Franz von Wieser, "Die Karte des Bartolomeo Columbo über die vierte Reise des Admirals," *Mitteilungen des Instituts für Österreichische Geschichtsforschung, Ergänzungsband* 4 (1893), pp. 488–498, which article is summarized (with illustrations of the maps) by George C. Hurlbut, "Geographical Notes," *Journal of the American Geographical Society of New York* 26 (1894), pp. 252–257. Also see John Bigelow, "The So-Called Bartholomew Columbus Map of 1506," *Geographical Review* 25.4 (1935), pp. 643–656, and Roberto Almagià, "Intorno a quattro codici fiorintini e ad uno ferrarese dell'erudito veneziano Alessandro Zorzi," *La Bibliofilía* 38 (1936), pp. 313–347, esp. 322–331; this article is reprinted in Almagià's *Scritti geografici* (Rome: Edizioni cremonese, 1961), pp. 447–468; the sketch maps are also illustrated in Nordenskiöld, *Periplus* (see note 40 in Chap. 5), pp. 167–169, figs. 79–81. For additional discussion of the sketch maps, see George E. Nunn, "The Three Maplets Attributed to Bartholomew Columbus," *Imago Mundi* 9 (1952), pp. 12–22.

[24] See on the place names from Columbus's Fourth Voyage on Zorzi's sketch map and see George E. Nunn, *World Map of Francesco Roselli Drawn on an Oval Projection and Printed from a Woodcut Supplementing the Fifteenth Century Maps in the Second Edition of the Isolario of Bartolomeo dali Sonetti, Printed in Italy Anno Domini MDXXXII* (Philadelphia: Press of John T. Palmer Co., 1928), pp. 8–12 and 19–23, and Nunn's "The Three Maplets Attributed to Bartholomew Columbus" (see note 23 in Chap. 6), p. 21.

[25] Skelton, "World Map by Henricus Martellus Germanus, c. 1489, at Berne" (see note 2 in Front Matter), p. 16.

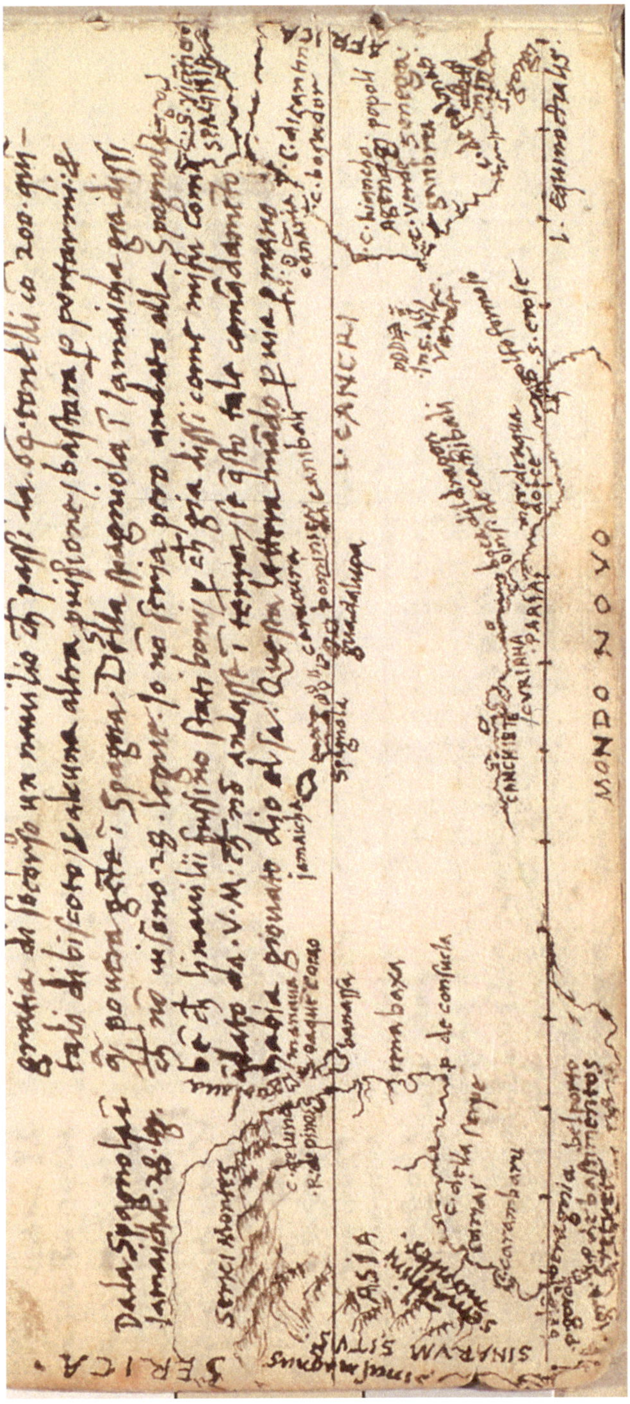

Fig. 6.8 Sketch map by Alessandro Zorzi (c. 1506) showing the influence of Martellus in the shape of eastern Asia (Florence, BNCF, Banco Rari 234, f. 60v). By concession of the Ministero dei Beni e delle Attività C.lturali e del Turismo/Biblioteca Nazionale Centrale, Firenze. (Further reproduction in any medium is forbidden)

passage—which is exactly the position ascribed to the island by Martellus on the Yale map. It is worth citing the relevant passage in full[26]:

> *yten si saben que el dicho martin alonso pincon dió aviso al dicho almirante don Cristobal colon dela tyerra e lo platycó conél por la escritura susodicha la qual se dyxo que hera sentencia del tienpo de salomon que resaba navegara por el mar mediterráneo fasta el fin despaña e allí al poniente del sol entre el norte e el medio dia por vía tenperada fasta noventa e cinco grados del camyno que fallaría una tyerra de sypanso la qual es tan fértyl y abundosa e con la su grandeza sojusgara a áfrica e europa.*

And an English translation[27]:

> *Also, if they know, etc., that Marín Alonso Pinzón advised the admiral don Cristóbal Colón of the land and discussed it with him using the abovementioned writing which he said was a maxim of the time of Solomon that said if you will navigate through the Mediterranean to the end of Spain and there towards the setting of the sun between north and south by a tempered path as far as ninety-five degrees of the way you will find the land of Cipangu which is very fertile and abundant and with its greatness will subjugate Africa and Europe.*

Skelton tries to interpret the reference "between north and south" as referring to the north-and-south orientation of Cipangu on the Yale Martellus map, but that is pushing too far. The document that Pinzón saw is certainly very strange—King Solomon is generally thought to have lived in the tenth century BC, long before anyone was measuring longitude, and Japan was certainly not known in the Mediterranean basin at that time—but the coincidence of position ascribed to Japan in this alleged maxim from the time of Solomon and on the Yale Martellus map is quite remarkable and suggests that a map by Martellus similar to the Yale Martellus map had been in circulation in Rome.

In addition, Skelton[28] and following in his footsteps, Randles,[29] have argued that when Columbus made landfall in the Caribbean on October 11, 1492, and immediately started looking for the island of Cipangu (Japan), his conviction that Japan must be near[30] was based on an almost

---

[26] The passage is from *Colección de documentos ineditos relativos al descubrimiento, conquista y organización de las antiguas posesiones españolas de ultramar, segunda serie* (Madrid: Est. tip. 'Sucesores de Rivadeneyra,' 1885–1932), vol. 8 (= *De los pleitos de Colón*, vol. 2), no. 97, for August 11, 1515, pp. 121–143, at 126, paragraph 12.

[27] William D. Phillips, Mark D. Johnston, and Anne Marie Wolf, *Testimonies from the Columbian Lawsuits* (Turnhout: Brepols, 2000), p. 173 (English) and 391 (Spanish).

[28] Skelton, "World Map by Henricus Martellus Germanus, c. 1489, at Berne" (see note 2 in Front Matter), p. 16.

[29] W. G. L. Randles, "La Cartographie de l'Atlantique à la veille du Voyage de Christophe Colomb," in *Actas do II Colóquio Internacional de História da Madeira, Funchal, setembro de 1989* (Funchal: Comissão para as Comemorações dos Descobrimentos Portugueses, 1990), pp. 925–935, esp. 932–934; the article is reprinted in his *Geography, Cartography and Nautical Science in the Renaissance: The Impact of the Great Discoveries* (Aldershot, Great Britain; and Burlington, VT: Ashgate Variorum, 2000). Also see W. G. L. Randles, "The Evaluation of Columbus' 'India' Project by Portuguese and Spanish Cosmographers in the Light of the Geographical Science of the Period," *Imago Mundi* 42 (1990), pp. 50–64, also reprinted in his *Geography, Cartography and Nautical Science in the Renaissance*.

[30] Christopher Columbus, *The Journal: Account of the First Voyage and Discovery of the Indies*, ed. Paolo Emilio Taviani and Consuelo Varela, trans. Marc A. Beckwith and Luciano F. Farina (Rome: Istituto poligrafico e Zecca dello Stato, Libreria dello Stato, 1992), vol. 1, pp. 46–47. Columbus also recorded his expectation of finding or belief that he had found Cipango on October 6 (pp. 34–35),

perfect coincidence between the distance he had traveled west and the distance indicated between Lisbon and Japan on the Yale Martellus map. The Yale Martellus map is the only surviving fifteenth-century map that assigns a longitude to Japan, and it is very tempting indeed to think that Columbus's expectations about the distance he would have to sail had been shaped by that map, particularly given the other indications that he had seen a map like it. However, the claims of an essentially perfect coincidence between the two distances in question are not well founded and require examination and recalculation.

Skelton writes:

> *When landfall was made on 11 October, Columbus's Journal shows a distance sailed of 1123 leagues, or about 89.5°, on a westerly magnetic course from the Canaries; and he at once began the search for Cipangu, which he believed to be not far off. The Berne map [i.e. the Yale Martellus map] lays down Cipangu between 90° and 99°W (of the Canaries)....*

Unfortunately Skelton does not explain how he concluded that the 1123 leagues Columbus had covered from the Canary Islands were equivalent to 89.5° of longitude. Randles offers more detail. He begins with a figure of 1142 leagues, which is what Columbus calculated was the distance he had traveled from the Canaries to reach Cuba—which he thought was Japan. Several passages in his journal indicate that Columbus believed that he was at 42° N, though he was in fact at about 21° N. Randles and apparently also Skelton accept the erroneous figure of 42° N,[31] although is most likely the result of a mistake by the copyist of the journal than an error by Columbus in determining his latitude.[32] Columbus explicitly indicated

October 21 (pp. 70–71), October 23 (pp. 72–73), October 24 (pp. 74–75), October 26 (pp. 76–77), December 24 (pp. 198–201), December 26 (pp. 208–209), and January 4, 1493 (pp. 226–227).

[31] Skelton's and Randle's acceptance of the journal's indication that Columbus was at 42° N when he found land is difficult to accept in the context of an attempt to demonstrate that Columbus's geographical ideas were influenced by the Yale Martellus map. If Columbus thought he was at 42° N when he made land in the Caribbean, why did he not head south to find Japan, whose northern tip according to Martellus was at about 30° N?

[32] See Columbus, *The Journal: Account of the First Voyage and Discovery of the Indies* (see note 30 in Chap. 6), entries for October 30, November 2, and November 21, 1492, on pp. 84–85, 88–89, and 114–115, respectively. In the last case, Bartolomé de las Casas, who transcribed the only surviving copy we have of the journal, recognized that the figure of 42° was suspect and added "if the manuscript from which I made this copy was not altered." Certainly navigators did make errors with regard to latitude: Evelyn Edson notes that Columbus reports that Bartolomeu Dias estimated that the Cape of Good Hope was at 45°S rather than 35°S, an error of 10°: see Edson, *The World Map, 1300–1492* (see note 7 in Chap. 1), pp. 208–209, citing Columbus's postil in his copy of d'Ailly's *Ymago mundi*, transcribed by Buron in his edition of the text. See Pierre d'Ailly, *Ymago mundi*, ed. Edmond Buron (Paris: Maisonneuve frères, 1930), vol. 1, pp. 206–209. With regard to Columbus's error during his First Voyage, Alberto Magnaghi suggested that Columbus falsified the latitude of Cuba and other discoveries in the Caribbean in order to deceive the Portuguese in "I presunti errori che vengono attribuiti a Colombo nella determinazione delle latitudini," *Bollettino della Reale Società Geografica Italiana*, series 6, vol. 5 = 64 [i.e. 65] (1928), pp. 459–494 and 553–582, esp. 490–493. However, J. A. Williamson's reply to Magnaghi, "The Early Falsification of West Indian Latitudes: Review," *Geographical Journal* 75.3 (1930), pp. 263–265, in which he suggests that the error was due (as Las Casas had suggested) to the copyist of the journal, is more plausible. Magnaghi replied to Williamson in "Ancora dei pretesi errori che vengono attribuiti a Colombo nella determinazione delle latitudini," *Bollettino della Reale Società Geografica Italiana*, series 6, vol. 7 (1930), pp. 497–515.

that he believed that there were 56 2/3 miles per degree of longitude at the equator,[33] but Randles claims that it is "exceedingly doubtful" that Columbus ever applied this belief in practical cartography,[34] and instead Randles uses for his calculations the more common figure of 66 2/3 miles per degree of longitude at the equator.

Randles multiplies 66 2/3 miles per degree (at the equator) by the cosine of 42° (the latitude Columbus believed himself to be at) to arrive at 50 miles per degree at a latitude of 42°. Columbus's 1142 leagues are equivalent to 4568 miles, and dividing this figure by the 50 miles per degree, one obtains 91.33°—a remarkable coincidence with the position of Japan on the Yale Martellus map between 90° and 100° west of the Canaries. This, it is claimed, is why Columbus immediately began searching for Japan when he made land in the Caribbean.

However, Skelton's implicit and Randles's explicit assumption that Columbus would calculate using 66 2/3 miles per degree rather than his professed value of 56 2/3 miles per degree is questionable at best. In the interest of seeing what Columbus's expectations were, it seems much more sensible to run the calculation using Columbus's preferred value. It also seems sensible to use a more realistic value for Columbus's latitude rather than the 42° N indicated in the manuscript of Columbus's journal that Las Casas was consulting—which figure Las Casas himself, who was no navigator, questioned. If we run the calculation using Columbus's figure of 56 2/3 miles per degree at the equator and Columbus's true latitude of about 21° N, we obtain a figure of 52.9 miles per degree at 21° N, which gives a location of 86.4° west of the Canaries. To a man who had seen Martellus map (which places the eastern coast of Japan at 90° west of the Canaries), and who calculated that he was 86.4° west of the Canaries, it would make perfect sense to start searching for Japan. Thus there is indeed a very good agreement between Columbus's expectations and the geography of the Yale Martellus map.[35]

I believe I have presented enough evidence to show that the Yale Martellus map accords very well with what Columbus believed about the

[33] See, for example, Columbus, *Select Letters* (see note 20 in Chap. 6), pp. 181–182; for a good discussion of Columbus's statements that a degree of longitude at the equator measured 56 2/3 miles, see George E. Nunn, "The Determination of the Length of a Terrestrial Degree by Columbus," in his *The Geographical Conceptions of Columbus: A Critical Consideration of Four Problems* (New York: American Geographical Society, 1924), pp. 1–30, and George E. Nunn, "The Imago Mundi and Columbus," *American Historical Review* 40.4 (1935), pp. 646–661, esp. 654–657.

[34] Randles, "The Evaluation of Columbus' 'India' Project" (see note 29 in Chap. 6), p. 60 in the *Imago Mundi* article, and p. 22 in the reprint in *Geography, Cartography, and Natural Science.*

[35] One question that I have not addressed in the preceding is the length of the league that Columbus was using. James E. Kelley, Jr., "In the Wake of Columbus on a Portolan Chart," *Terrae Incognitae* 15 (1983), pp. 77–111, suggests that he was using the Italian league of 2.67 miles, and it is worth running through the calculations with this figure. In this case, Columbus's 1142 leagues would be 3049 miles, and calculating for a position at 42° N and using Columbus's figure of 56 2/3 miles per degree of longitude at the equator, we obtain a figure of 42.1 miles per degree at 42° N, so Columbus would have thought that he was 72.4° west of the Canaries when he was at Cuba. Using the more realistic position of 21° N, at which using Columbus's figure of 56 2/3 miles per degree of longitude at the equator, there would be 52.9 miles per degree, Columbus would have concluded that he was 57.6° west of the Canaries. It is difficult to believe that Columbus would have been quite as expectant of encountering Japan soon if he had been calculating using a league of 2.67 miles.

lands he hoped to find, particularly regarding the position and shape of Japan and mainland Asia. The question that remains is, were Columbus's expectations about Asia shaped by his consultation of the Yale Martellus map (or another very similar map produced by Martellus), or were both Columbus's and Martellus's views of the geography of Asia shaped by an earlier map to which both had access? The only known candidate for such an earlier map is the Toscanelli map, that is, the map that the Italian mathematician and cosmographer Paolo dal Pozzo Toscanelli (1397–1482)[36] supposedly sent together with a letter dated June 25, 1474, to Fernão Martins, a priest in Lisbon—which map no longer survives. The letter outlined Toscanelli's plan to sail west from Europe to reach the Spice Islands and Asia. Fernão Martins was said to have delivered the letter to King Afonso V of Portugal; Columbus is said to have heard about Toscanelli's plan, to have written to him asking for copies of the letter and of the map, and to have received both from him.

Toscanelli's letter and map are highly controversial; the most heated debate took place just after the turn of the twentieth century.[37] Many more recent scholars seem to have simply dismissed the doubts that have been raised about the authenticity of the letter and map and do not address the issue—as if at some point those doubts had been resolved, but I know of no convincing resolution of most of them.[38] The main pieces of evidence

[36]On Toscanelli, see Gustavo Uzielli, *La vita e i tempi di Paolo dal Pozzo Toscanelli* (Rome: Ministero della pubblica istruzione, 1894); Vespasiano da Bisticci, *Renaissance Princes, Popes, and Prelates: The Vespasiano Memoirs, Lives of Illustrious Men of the XVth Century*, trans. William George and Emily Waters (New York: Harper & Row, 1963), pp. 423–424; G. Abetti, "Toscanelli dal Pozzo, Paolo," in Charles Coulston Gillispie, ed., *Dictionary of Scientific Biography* (New York: Scribner, 1970–1980), vol. 13, pp. 440–441; Jane L. Jervis, *Cometary Theory in Fifteenth-Century Europe* (Dordrecht and Boston: D. Reidel, 1985), pp. 43–85 and 162–169, a particularly good account; Leonardo Rombai, "Paolo dal Pozzo Toscanelli (1397–1482) umanista e cosmografo," *Rivista geografica italiana* 100.1 (1993), pp. 133–158; and Eugenio Garin, "Ritratto di Paolo dal Pozzo Toscanelli," in his *Ritratti di umanisti: sette protagonisti del Rinascimento* (Milan: Tascabili Bompiani, 2001), pp. 40–67.

[37]The early twentieth-century bibliography on the Toscanelli controversy is gathered by Henry Vignaud, *Bibliografia della polemica concernente Paolo Toscanelli e Cristoforo Colombo* (Naples: Tip. ed. Cav. A. Tocco-Salvietti, 1905). The most important work on Toscanelli's letter and map is Henry Vignaud, *Toscanelli and Columbus: The Letter and Chart of Toscanelli on the Route to the Indies by Way of the West, Sent in 1474 to the Portuguese Fernam Martins, and Later on to Christopher Columbus* (London, Sands & Co., 1902); John Boyd Thacher has helpful chapters titled "The Letter of Toscanelli in the Spanish, Latin, and Italian Versions" and "The Import of the Letter" in his *Christopher Columbus: His Life, His Work, his Remains as Revealed by Original Printed and Manuscript Records* (New York and London: G. P. Putnam's Sons and The Knickerbocker Press, 1903–1904), vol. 1, pp. 301–316 and 317–324, respectively. A more recent study of the matter may be found in Miles H. Davidson, "The Toscanelli Letters: A Dubious Influence on Columbus," *Colonial Latin American Historical Review* 5.3 (1996), pp. 287–310, parts of which appear in his book *Columbus Then and Now: A Life Reexamined* (Norman: University of Oklahoma Press, 1997), pp. 50–61.

[38]Clements Markham disputed some of Vignaud's conclusions in an article titled "New Theories on Columbus" in the *Times Literary Supplement* (London) 43 (November 7, 1902), pp. 330–331, and subsequently wrote a more elaborate expression of his views in a letter to Vignaud. Vignaud published Markham's longer letter and his own (effective) reply in Clements R. Markham and Henry Vignaud, *Toscanelli and Columbus: A Letter from Sir Clements R. Markham... and a Reply from Mr. Henry Vignaud...* (London: Sands & Co., 1903). Vignaud replied to criticisms by Hermann Wagner and Carlo Errara in *The Columbian Tradition on the Discovery of America and of the Part Played Therein by the Astronomer Toscanelli: A Memoir Addressed to the Professors Hermann Wagner of the University of Göttingen and Carlo Errara of Bologna* (Oxford: The Clarendon Press, 1920). The dispute may be traced further in the bibliography cited at the beginning of the preceding note.

favoring the genuineness of the letter and map and the reality of their influence on Columbus are the following. First, Toscanelli's letter is copied by hand in Latin in what some have claimed to be Columbus's handwriting[39] in what is indubitably Columbus's copy of the 1477 edition of Pope Pius II's *Historia rerum ubique gestarum*.[40] In addition, the text of the letter from Toscanelli to Columbus (i.e., the letter to Martins preceded by a paragraph by Toscanelli addressed to Columbus in which he says that he is including a copy of the map)[41] is supplied in Italian by Fernando Colón in Chapter 8 of his biography of his father, first published in 1571,[42] and in Spanish by Bartolomé de las Casas in Book 1, Chapter 12, of his *Historia de las Indias*, which was first published in 1875.[43] Both authors claim that the letter influenced Columbus in the formulation of his plan to sail west,[44]

[39]For an argument that the handwriting is that of Columbus, see Thacher, "The Handwriting—Continued," in his *Christopher Columbus* (see note 37 in Chap. 6), vol. 3, pp. 454–488, esp. 461–474 and 485.

[40]Columbus's copy of the *Historia rerum ubique gestarum*, which has many of his annotations in addition to the handwritten copy of the letter, has been reproduced in facsimile as Pope Pius II, *Historia rerum: cuyo original se encuentra en la Biblioteca Colombina de Sevilla* (Madrid: Testimonio Compañía Editorial, 1991), with a volume of commentary by Juan Pérez de Tudela y Bueso, *La Historia rerum ubique gestarum del papa Pio II y el descubrimiento de América* (Madrid: Testimonio Compañía Editorial, 1993). There is a facsimile of the letter and transcription in Cesare De Lollis, *Autografi*, in the *Raccolta di documenti e studi pubblicati dalla R. Commissione colombiana* (Rome: Ministero della pubblica istruzione, 1892–1896), part 1, v. 3, plates 72 and 73, which are reproduced in Vignaud, *Toscanelli and Columbus* (see note 37 in Chap. 6), Appendix G, between pages 336 and 337; other reproductions are listed by Vignaud on his pp. 13–14 and other translations into English and French on his pp. 16–18. The Latin text of the letter is supplied by Vignaud, *Toscanelli and Columbus*, Appendix B, pp. 293–303, and translated into English by Vignaud, *Toscanelli and Columbus*, Appendix A, pp. 275–292; the Latin text and an English translation are provided by Thacher, *Christopher Columbus* (see note 37 in Chap. 6), vol. 1, pp. 308–312, and an English translation by Davidson, "The Toscanelli Letters," pp. 294–297. The Latin text of the letter is supplied with a Spanish translation and brief introduction in Juan Gil, "La correspondencia con Toscanelli," and in Juan Gil and Consuelo Varela, eds., *Cartas de particulares a Colón y relaciones coetáneas* (Madrid: Alianza Editorial, 1984), pp. 129–141.

[41]Vignaud edits and translates Fernando and Las Casas's versions of this introductory paragraph in *Toscanelli and Columbus* (see note 37 in Chap. 6), Appendix E, pp. 319–321.

[42]See Fernando Colón, *The History of the Life and Deeds of the Admiral Don Christopher Columbus, Attributed to his Son Fernando Colón*, ed. Ilaria Caraci Luzzana, trans. Geoffrey Symcox and Blair Sullivan (Turnhout: Brepols, 2004), chapter 8, pp. 42–44 (English) and 245–247 (Italian). The Italian text of the letter together with an English translation is also supplied by Vignaud, *Toscanelli and Columbus* (see note 37 in Chap. 6), Appendix D, pp. 311–317, and by Thacher, *Christopher Columbus* (see note 37 in Chap. 6), vol. 1, pp. 312–316. Other reproductions are listed by Vignaud, p. 15.

[43]See Bartolomé de las Casas, *Obras completas*, ed. Paulino Castañeda Delgado (Madrid: Alianza, 1988–1998), vol. 3 = *Historia de las Indias*, vol. 1, edited by Miguel Angel Medina, Jesús Angel Barreda, and Isacio Pérez Fernández, Book 1, chapter 12, pp. 398–402. The Spanish text of the letter together with an English translation is also supplied by Vignaud, *Toscanelli and Columbus* (see note 37 in Chap. 6), Appendix C, pp. 305–310, and by Thacher, *Christopher Columbus* (see note 37 in Chap. 6), vol. 1, pp. 302–307. Also see Vignaud, p. 15.

[44]Fernando Colón writes: "...he was also influenced by a Master Paolo, physician to Master Domenico, a Florentine, a contemporary of the admiral's, who played a role in encouraging him to undertake this voyage. This Master Paolo was a friend of Fernão Martins, a canon of Lisbon, and they wrote each other letters about the voyages to the country of Guinea that were being made at the time of King Alfonso of Portugal and the possibility of other voyages to the western regions. This came to the attention of the admiral, who was extremely interested in these things, and immediately, through the agency of Lorenzo Girardi, a Florentine who was residing in Lisbon, he wrote

and both also supply the translated text of another letter that Toscanelli allegedly sent to Columbus in which he congratulates Columbus for his desire to undertake the voyage westward.[45] In addition, Las Casas says that he had Toscanelli's map in his possession when he was writing his *Historia de las Indias*.[46] And although there are various problems involved in determining Columbus's thoughts and calculations involving the location of Japan (as indicated just above in my discussion of the location of Japan on the Martellus map), he did find land at least approximately where the Toscanelli letter, and no doubt the map, indicated that Japan should be.[47]

This *prima facie* evidence of the genuineness of the letter and map seems quite solid, but strong arguments have been raised against them, particularly by Vignaud and more recently by Davidson. Among the arguments put forward by Vignaud are the following. He notes that there is no sign that Toscanelli's letter, which was supposedly sent to the king of Portugal, ever made any impression on the Portuguese court and that his plan for sailing west was unknown in Italy.[48] More tellingly, Vignaud notes that Columbus never mentions or cites Toscanelli in any of his writings and in particular never says that the map he had with him on his First Voyage was by Toscanelli.[49]

---

about this to Master Paolo and sent him a small sphere on which he indicated his plan. Master Paolo replied to him in Latin, which translated into the vernacular reads as follows." See Colón, *The History of the Life and Deeds of the Admiral Don Christopher Columbus* (see note 42 in Chap. 6), Chapter 7, pp. 42 and 245.

[45]This so-called second letter appears in Las Casas, *Historia de las Indias*, Book 1, Chapter 12, and Colón, *The History of the Life and Deeds of the Admiral Don Christopher Columbus* (see note 42 in Chap. 6), Chapter 8. Vignaud edits and translates Las Casas's and Fernando's versions of this other letter in *Toscanelli and Columbus* (see note 37 in Chap. 6), Appendix E, pp. 322–326.

[46]In Book 1, Chapter 12 of the *Historia de las Indias*, Las Casas writes: "…la carta de marear que le invió, yo, que esta historia escribo, tengo en mi poder y della se hará más mencion abaxo. Mucho ánimo le puso con ella, y, sino supiera más, por ella y por las cosas de suso traídas, sin duda del todo se moviera; y así creo que todo su viaje sobre esta carta fundó, pero aún más se lo quiso nuestro Señor declarar, como se verá," that is, "The sea-chart he sent him is now in my possession as I write this account and we shall have more to say about it, too, later. It filled him with enthusiasm and even had he known nothing up to that point, it would, together with the other reasons we have mentioned, have been sufficient to spur him to action, and I believe he based the whole of his voyage on this chart—though our Lord, as we shall see, still wished to reveal yet more to him." The Spanish text is from Bartolomé de las Casas, *Obras completas*, ed. Paulino Castañeda Delgado (see note 43 in Chap. 6), vol. 3, p. 402, and the translation from Bartolomé de las Casas, *Las Casas on Columbus: Background and the Second and Fourth Voyages*, ed. Nigel Griffin (Turnhout: Brepols, 1999), p. 36. Las Casas talks about the chart more in Book 1, Chapters 38 and 43 of the *Historia de las Indias*.

[47]See Vignaud, *Toscanelli and Columbus* (see note 37 in Chap. 6), pp. 207–211.

[48]See Vignaud, *Toscanelli and Columbus* (see note 37 in Chap. 6), pp. 36–38 and 39–43. Vignaud also argued that Fernam Martins, the supposed recipient of the original letter from Toscanelli, was totally unknown, and therefore his existence is dubious (pp. 38–39). However, António Domingues de Sousa Costa presents excellent evidence that Fernam Martins de Reriz did indeed exist in his article "Cristóvão Colombo e conégo de Lisboa Fernando Martins de Reriz, destinatário da carta de Paúlo Toscanelli sobre os descobrimentos marítimos," *Antonianum* 65 (1990), pp. 187–276, esp. 242–250. Also see Eric C. Apfelstadt, "Christopher Columbus, Paolo del Pozzo Toscanelli and Fernão de Roriz: New Evidence for a Florentine Connection," *Nuncius* 7 (1992), pp. 69–80, who presents evidence that Fernando de Roriz was in Florence in 1459.

[49]Vignaud, *Toscanelli and Columbus* (see note 37 in Chap. 6), pp. 47–51 and 222–225.

Vignaud also points out a significant anachronism in the letter. The author of the letter writes[50]:

> *Cum tecum allias locutus sum de breviori via ad loca aromatum per maritimam navigacionem quam sit ea quam facitis per guineam querit nunc serenissimus rex a me quandam declaracionem ymo potius ad occulum ostensionem vt etiam mediocriter doti illam viam caperent et intelligerent.*
>
> *I have spoken with you elsewhere concerning a shorter way of going by sea to the lands of spices, than that which you are making by Guinea; the most serene King now wishes that I should give some explanation thereof, or rather that I should so set it before the eyes of all, that even those who are but moderately learned might perceive that way and understand it.*

The problem is, as Vignaud points out, that there is no evidence that in 1474, the purported date of the letter, the Portuguese were seeking to access the land of spices (i.e., the islands of the Indian Ocean) by sailing around Africa; instead, they were seeking the India of Prester John, in order to establish ties with that Christian monarch.[51] In fact, it is easy to corroborate Vignaud's argument with additional evidence. First, Diogo Cão on voyages down the coast of Africa in 1482–1484 and 1485–1486 left *padrões* or inscribed pillars indicating how far he had traveled and when, and the inscriptions on these pillars mention a goal only of exploring territory, not of seeking a route to the Spice Islands of the Indian Ocean.[52] And the purpose of the voyage of Bartolomeu Dias, who was the first European to round the southern tip of Africa on his voyage of 1487–1488, was to search for Prester John.[53] In addition, although Marco Polo, John Mandeville, and Nicolò dei Conti had mentioned spices on various islands in the Indian Ocean, it is not until 1487, more than a dozen years after Toscanelli's letter was allegedly composed, that we have good evidence of Portuguese's interest in reaching the spices of the East: King John II of Portugal sent Pero da Covilhã on a mission to the East, and one of his goals

[50]The Latin and English are from Vignaud, *Toscanelli and Columbus* (see note 37 in Chap. 6), pp. 294–295 and 277–278, respectively.

[51]Vignaud, *Toscanelli and Columbus* (see note 37 in Chap. 6), pp. 52–65. C. Raymond Beazley, "Toscanelli and Columbus," *The Guardian*, November 19, 1902, takes issue with Vignaud's claims on this subject, but Vignaud very ably replies in *Toscanelli and Columbus: Letters to Sir Clements R. Markham... and to C. Raymond Beazley, M. A.* (London: Sands & Co., 1903), pp. 26–31. For discussion of the early development of the spice trade in Portugal, see Donald F. Lach, "The Spice Trade," in his *Asia in the Making of Europe*, vol. 1, The Century of Discovery (Chicago: University of Chicago Press, 1965), Book 1, pp. 91–115, esp. 91–103. On the later development of trade along the route from Portugal around Africa to India, see T. Bentley Duncan, "Navigation Between Portugal and Asia in the Sixteenth and Seventeenth Centuries," in Cyriac K. Pullapilly and Edwin J. Van Kley, eds., *Asia and the West: Encounters and Exchanges from the Age of Explorations* (Notre Dame: Cross Cultural Publications, 1986), pp. 3–25; reprinted in Om Prakash, ed., *European Commercial Expansion in Early Modern Asia* (Aldershot, England; Brookfield, VT: Variorum, 1997).

[52]See E. G. Ravenstein, "The Voyages of Diogo Cão and Bartholomeu Dias, 1482–88," *The Geographical Journal* 16.6 (1900), pp. 625–655, esp. 628 and 634. On the development of Portuguese goals for voyages to the East, also see Luís Filipe F. R. Thomaz, "O Projecto Imperial Joanino (Tentativa de interpretação global da política ultramarina de D. João II)," in *Congresso Internacional Bartolomeu Dias e a sua Epoca: Actas* (Porto: Universidade do Porto and Comissão Nacional para as Comemorações dos Descobrimentos Portugueses, 1989), vol. 1, pp. 81–98, esp. 87–91.

[53]Ravenstein, "The Voyages of Diogo Cão and Bartholomeu Dias, 1482–88" (see note 52 in Chap. 6), p. 638.

was to determine where exactly the spices came from.[54] So it is quite certain that in 1474 the goal of sailing around Africa for the purpose of reaching the Spice Islands had not been clearly formulated.

Vignaud points out another serious problem with the letter.[55] The author claims to have spoken at length with an ambassador from Cathay who came to Rome during the papacy of Eugene IV, who held office from 1431 to 1447[56]:

> *etiam tempore Eugenii venit unus ad eugenium qui de benivolentia magna erga christianos afirmabat et ego secum longo sermone locutus sum de multis, de magnitudine edificiorum regalium et de magnitudine fluvium in latitudine et longitudine mirabili et de multitudine civitatum in ripis fluuium ut in uno flumine 200 e civitates sint constitute et pontes marmorei magne latitudinis et longitudinis undique colonpnis ornati.*
>
> *In the time of (Pope) Eugenius, also, one came to Eugenius and spoke of (their) great goodwill towards Christians. And I held speech with him for a long time on many things, on the greatness of the royal buildings, and on the greatness of the rivers of wondrous breadth and length, and on the multitude of cities on the banks of the rivers; and how on one river there are established about two hundred cities, and marble bridges of great breadth and length adorned with columns on every side.*

This passage well exemplifies a point that Vignaud makes about the whole letter: that it does not seem to be the product of the searching intellect of an eminent humanist, but instead focuses on "details as childish as exaggerated." There is very little science in the letter, only a glowing description of the riches of the East. Moreover, Vignaud notes that although the author claims to have recent information from an ambassador from Cathay, in fact the letter contains no new information and employs terms (Cathay, Great Khan, Mangi, Zayton, Quinsay) that were no longer in use in China in the fifteenth century. These terms were in use while the Mongols controlled China but had fallen by the wayside with the advent of the Ming Dynasty in 1386. This is very powerful evidence indeed that the letter is not what it claims to be.

Davidson in his discussion of the Toscanelli letters adds some details to corroborate Vignaud's conclusions. Of these, the most important seem to me to be his citation of Sumien's opinion that the Latin version of the letter (rather than the Spanish or Italian) is the original while pointing out that the Latin is far beneath the level of a famous humanist[57] and his supplying of evidence that at the alleged date of Toscanelli's letter, the

[54]See C. F. Beckingham, "The Travels of Pero da Covilhã and Their Significance," *Actas, Congresso Internacional da História dos Descobrimentos* (Lisbon: Comissão Executiva das Comemorações do V Centenário da Morte do Infante D. Henrique, 1961), vol. 3, pp. 1–14, reprinted in his *Between Islam and Christendom: Travellers, Facts, and Legends in the Middle Ages and the Renaissance* (London: Variorum Reprints, 1983).

[55]Vignaud, *Toscanelli and Columbus* (see note 37 in Chap. 6), pp. 67–70.

[56]The Latin text and translation are from Vignaud, *Toscanelli and Columbus* (see note 37 in Chap. 6), pp. 299–300 and 284–285, respectively.

[57]See Miles H. Davidson, "The Toscanelli Letters: A Dubious Influence on Columbus," *Colonial Latin American Historical Review* 5.3 (1996), pp. 287–310, at 290 and 292, and Norbert Sumien, *La correspondance du savant florentin Paolo dal Pozzo Toscanelli avec Christophe Colombe* (Paris: Société d'Éditions Géographiques, Maritimes et Coloniales, 1927).

Portuguese were seeking slaves and gold in their explorations down the coast of Africa rather than a route to the Spice Islands.[58]

There are some important points about the letter that should be added to Vignaud's and Davidson's arguments. Vignaud's point about the intellectual level of the letter being well below what we would expect from an eminent humanist is well taken, but can and should be further developed. An eminent humanist, in discussing the possibility of sailing across the Atlantic to reach Asia, would certainly cite some of the pronouncements[59] of Aristotle (*De caelo* 2.14),[60] Seneca (*Questiones naturales*, preface),[61] Pliny (*Naturalis historia* 2.67.169), Roger Bacon,[62] and Pierre d'Ailly[63] on the narrowness of the Atlantic, and yet the author of the letter says not a word on this subject. An eminent Florentine humanist of the year 1474 would surely be well acquainted with Ptolemy's *Geography* and

---

[58] Davidson, "The Toscanelli Letters" (see note 57 in Chap. 6) pp. 297–299.

[59] Most of these authorities and some others as well are cited by Fernando Colón in his biography of his father when he discusses his father's motivations for sailing across the Atlantic: see Colón, *The History of the Life and Deeds of the Admiral Don Christopher Columbus* (see note 42 in Chap. 6), Chapter 7, pp. 40–42.

[60] For the discussion of the relevant passage in Aristotle's *De caelo* 2.14, see Charles Jourdain, "De l'influence d'Aristote et de ses interprètes sur la découverte du Nouveau-Monde," *Journal général de l'instruction publique* 30.63 (August 7, 1861), pp. 495–498, and 30.64 (August 10, 1861), 504–506, reprinted in his *Excursions historiques et philosophiques à travers le Moyen Age* (Paris, 1888), pp. 587–616. Fernando Colón (see the previous note) remarks that Averroes cites this passage in Aristotle with approval, and the passage in Averroes's commentary may be consulted in *Averrois Cordubensis commentum magnum super libro De celo et mundo Aristotelis*, ed. Rüdiger Arnzen (Leuven, Belgium: Peeters, 2003), Book 2, section 111, in vol. 2, pp. 480–481.

[61] See Lucius Annaeus Seneca, *Physical Science in the Time of Nero: Being a Translation of the 'Quaestiones naturales' of Seneca*, trans. John Clarke (London: Macmillan and Co., Limited, 1910), p. 7. In the recent translation of Seneca's *Natural Questions* by Harry M. Hine (Chicago and London: The University of Chicago Press, 2010), the parts of the work are placed in what is generally agreed to have been their original order, and the passage in question appears in what is here called Book 1 (originally Book 7), printed following Book 7 (originally Book 6), on p. 138: "For what distance lies between the farthest coasts of Spain and the Indies? An interval of very few days, if a ship is driven by a favorable wind." For discussion of Columbus's use of this passage, see James Romm, "New World and 'novos orbes': Seneca in the Renaissance Debate over Ancient Knowledge of the Americas," in Wolfgang Haase and Meyer Reinhold, eds., *The Classical Tradition and the Americas*, vol. 1, part 1, *European Images of the Americas and the Classical Tradition* (Berlin and New York: De Gruyter, 1994), pp. 78–116, esp. 88–89. This same chapter discusses a passage from Seneca's *Medea* that is mentioned by Fernando Colón as foretelling his father's discoveries, which is also addressed in Diskin Clay, "Columbus' Senecan Prophecy," *American Journal of Philology* 113.4 (1992), pp. 617–620.

[62] See Roger Bacon, *The 'Opus majus' of Roger Bacon*, ed. John Henry Bridges (Oxford: Clarendon Press, 1897–1900), vol. 1, pp. 290–291 (Latin text), and Roger Bacon, *The Opus majus of Roger Bacon*, trans. Robert Burke (Philadelphia: University of Pennsylvania Press; and London: H. Milford, Oxford University Press, 1928), vol. 1, p. 535 (English translation).

[63] The relevant passages from Pierre d'Ailly's *Imago Mundi* (Chapters 8 and 49) may be consulted in Pierre d'Ailly, *Ymago mundi*, ed. Edmond Buron (Paris: Maisonneuve frères, 1930), vol. 1, pp. 206, 208, and 211, and vol. 2, pp. 424, 426, and 427. These passages are quoted in English by Arthur Percival Newton, ed., *Travel and Travellers of the Middle Ages* (London: K. Paul, Trench, Trubner & Co., Ltd.; New York: A. A. Knopf, 1926), pp. 17–18. Some of Columbus's postils or marginal notes on the width of the Atlantic in his copy of d'Ailly's book are supplied in the edition of the *Ymago mundi* just cited, vol. 1, p. 208; also see the facsimile edition of Columbus's copy of the book, published as Pierre d'Ailly, *Imago mundi* (Madrid: Testimonio Compañía Editorial, 1990), f. 13r, together with the accompanying translation by Antonio Ramírez de Verger, *Imago mundi del Cardenal Pedro d'Ailly y Juan Gerson* (Madrid: Testimonio Compañía Editorial, 1990), p. 66.

would be able to describe a graticule of meridians and parallels more elegantly and clearly than thus: "The straight lines, therefore, marked lengthwise in the chart, show the distances from east to west, but those which are transverse show the space from south to north"[64] and would speak of degrees of longitude, rather than the vague "spaces" used in the letter. A distinguished humanist interested in the spices of the orient, given the opportunity to talk to an ambassador from China, would certainly ask about the sources and cultivation of those spices, but the letter attributed to Toscanelli shows no signs of such an enquiry.[65] In addition, John Shipley has correctly noted that the text attributed to Toscanelli is "a rambling, digressive letter, wanting in logical arrangement.... far from being the production of a scholarly mind accustomed to logical sequence."[66] Shipley continues:

> *To support the charge of digression, we may adduce the fact that the Letter begins to describe the map, then proceeds to enlarge upon the countries that may be reached, and does not resume the description of the map until the Postscript.... In this Postscript the writer, after mentioning the 26 spaces between Lisbon and Quinsay, goes on to describe the latter city, then adds, several words further on, the remark that these 26 spaces are about a third of the whole circumference. Later on he again returns to the map and the 10 spaces between Antilia and Cipango, but he finishes his description of the latter island before remarking that these 10 spaces represent no great extent of unknown sea to be traversed.*

The way that the alleged Toscanelli letter is addressed also raises grave questions. Why would Toscanelli not send his letter and map directly to the king of Portugal? There is no clear advantage to sending them to an intermediary, but many possible advantages to sending them directly to the king. The chances of deriving some benefit from the proposal of this potentially enormously valuable idea would be far greater if the author had addressed himself directly to the king and had couched the letter in the appropriate terms. Indeed, this method of proceeding would have demonstrated more respect to the king and would seem likely to have obtained for the letter a more serious hearing, even if the author had no interest in deriving any material benefit from the proposal. It is worth contrasting Toscanelli's alleged manner of offering his idea to the king of Portugal with the procedure adopted by a man who actually did have the idea of sailing out across the Atlantic, namely, Columbus. It is true that

[64] See Vignaud, *Toscanelli and Columbus* (see note 37 in Chap. 6), p. 280 with footnote.

[65] It is worth remarking that the author of the so-called anonymous narrative of Cabral's voyage in the *Paesi novamente retrovati*, an important collection of travel narratives first published in 1507, Chapters 82 and 83, supplies a detailed account of the sources and prices of spices in Calicut, India. These chapters are translated into English by William Brooks Greenlee, *The Voyage of Pedro Álvares Cabral to Brazil and India from Contemporary Documents and Narratives* (London: The Hakluyt Society, 1938), pp. 53–94, at 91–94.

[66] John B. Shipley, "Notes on the So-Called Second Letter of Toscanelli, Supposed to Have Been Addressed to Christopher Columbus, and Its Bearing on the History of the So-Called First Letter," in F. W. Putnam, Franz Boaz, and M. H. Saville, eds., *International Congress of Americanists: Thirteenth Session Held in New York in 1902* (Easton, PA: Eschenbach Print Co., 1905), pp. 305–325, at 313–314. I do not agree with Shipley's other conclusions about the letter, but his description of its rambling nature (quoted here) is accurate.

Columbus was proposing to make the voyage himself, while the author of the letter was not, but Columbus took his idea directly and in person to the monarchs who might be able to finance it.[67]

The so-called second letter of Toscanelli purports to be Toscanelli's reply directly to Columbus, evidently in response to a letter in which Columbus had mentioned his interest in reaching Asia by sailing to the West.[68] Vignaud addresses the question of when the letter was supposedly written and points out that the letter is obviously a draft of the first letter, as it shares many phrases and ideas with the first letter, and the letter pretends to accompany the map (which the so-called first letter does as well). He concludes that the second letter is the remains in an earlier stage in the development of the forged correspondence: "The author of the fraud clearly began by making Toscanelli correspond directly with Columbus; then he substituted for this latter the unfindable Martins."[69] But really the strongest evidence that the second letter is a forgery—which makes it essentially certain that the first is as well—is its pure vacuousness. Here is the first half or so of the text[70]:

> *To Christopher Columbus, Paul, the physician, health:*
>
> *I have received thy letters with the things thou didst send me, and with them I received a great favour. I notice thy splendid and lofty desire to sail to the regions of the east by those of the west, as is shown by the chart which I send you, which would be better shown in the shape of a round sphere; it will please me greatly, should it be understood; and that not only is the said voyage possible, but it is sure and certain, and of honour and countless gain, and of the greatest renown among all Christians. But you will not be able to understand it thoroughly except with experience or discussion, as I have had most fully, and good and true information of mighty men and of great learning, who have come from the said regions here to the Court of Rome, and of other merchants who have long trafficked in those parts, men of great authority. So that when the said journey occurs, it will be to powerful kingdoms and most noble cities and provinces, most rich in all manner of things in great abundance and very necessary to us, as also in all kinds of spices in great quantity, and of jewels in the largest abundance.*

In this letter, part of an alleged correspondence between one of the foremost navigators of the day and one of the foremost geographers of the day about a bold plan whose success depended on the size and characteristics

[67] For a brief discussion of Columbus's negotiations with monarchs about his plan, see Samuel E. Morison, *The European Discovery of America*, vol. 2, *The Southern Voyages, A.D. 1492–1616* (New York: Oxford University Press, 1974), pp. 31–44. For an interesting discussion of the reconception of the ocean involved in Columbus's plan to sail west, see Thomas Goldstein, "Florentine Humanism and the Vision of the New World," in *Actas do Congresso Internacional de História dos Descobrimentos* (Lisbon: Comissão Executiva das comemoracões do V Centenário da Morte do Infante D. Henrique, 1961), vol. 4, pp. 195–208, and the same author's "Geography in Fifteenth-Century Florence," in John Parker, ed., *Merchants and Scholars: Essays in the History of Exploration and Trade* (Minneapolis: University of Minnesota Press, 1965), pp. 9–32, reprinted in Felipe Fernández-Armesto, ed., *The European Opportunity* (Aldershot, Great Britain, and Brookfield, VT: Variorum, 1995), pp. 1–22.

[68] For references on the second letter of Toscanelli, see note 45 in Chap. 6.

[69] See Vignaud, *Toscanelli and Columbus* (see note 37 in Chap. 6), pp. 31–35, 153, and 271, with the quotation coming from the last of these pages.

[70] This is Vignaud's English translation of the Spanish version in Las Casas, *Historia de las Indias*, Book 1, Chapter 12, *Toscanelli and Columbus* (see note 37 in Chap. 6), pp. 322–323.

of the ocean and the geography of Asia, there is no reference to or question about any specific geographical text or map (aside from the alleged Toscanelli map), no discussion of the length of a degree of longitude in miles, no remark about a specific island, and no mention of currents or winds. There is no hint of a disagreement about any detail, as there naturally would be in any discussion of a plan involving territory about which so little was known. There is not even a little humanistic flaunting of scholarship for its own sake.

Very little of Toscanelli's writings survives, but a manuscript in the Biblioteca Nazionale Centrale in Florence does contain some of his observations of comets.[71] Although the subject and genre of these notes is quite different than that of the alleged Toscanelli letter, it is worth citing some of the notes to show that Toscanelli was indeed a scientist interested in details and evaluating evidence, in contrast with the author of the letters attributed to him[72]:

> *apparuit comata* [sic] *stella mihi primo die dominico 4 octobris 1433 in die sancti Francisci prima hora noctis in imagine corone Adriane versus eius partem meridionalem, ut in figura patet. visa tamen fuit ab alio, quomodo ex quodam scivi die sabati precedente 3 octobris. die 5 octobris lune in secundo visa est extra Coronam in parte meridionali magis aliquantulum; hac die visa est magis clare* [clara], *prima enim die visa est satis plumbea. nescio an hoc erat quia aer non ita serenus fuit sicut hac die, et etiam cauda in obscuritate noctis videbatur longior. magnitudo eius erat sicut stellarum secunde magnitudinis sed non tante claritatis.*
>
> *The comet first appeared to me on Sunday, October 4, 1433 on the day of St. Francis, first hour of the night, in the constellation Corona Ariadne toward its southern part, as shown in the figure. However it was seen by others, as was made known to me, the preceding Saturday, October 3. On Monday, October 5, its second appearance, it was outside the Corona a little bit more on the southern side; that day it appeared brighter, for the first day it appeared rather leaden. I do not know whether this was because the air was not so clear. The tail was seen longer in the darkness of night. Its magnitude was of a second magnitude star but not as brilliant.*

The second letter attributed to Toscanelli cannot be genuine, must be a forgery, and ineluctably drags the other Toscanelli letter and the Toscanelli map down to the same category.

Who forged the two letters and the map and when (Vignaud and Davidson address those subjects) is of less importance to me than the fact that they are derivative products, leaving it vastly more likely that it was Martellus's work that influenced Columbus, Behaim, and no doubt the

---

[71] The manuscript is Florence, Biblioteca Nazionale Centrale, Banco Rari 30. For discussion of the folios with Toscanelli's comet observations, see Jane L. Jervis, "Toscanelli's Cometary Observations: Some New Evidence," *Annali dell'Istituto e Museo di storia della scienza di Firenze* 2.1 (1977), pp. 15–20, and Jane L. Jervis, *Cometary Theory in Fifteenth-Century Europe* (Dordrecht and Boston: D. Reidel, 1985), pp. 43–85.

[72] The Latin is from BNCF Banco Rari 30, f. 245v, transcribed by G. Celoria in Gustavo Uzielli, *La vita e i tempi di Paolo dal Pozzo Toscanelli* (Rome: Ministero della Pubblica Istruzione, 1894), p. 306, with corrections in Jervis, *Cometary Theory in Fifteenth-Century Europe*, p. 56, who also supplies the English translation cited here. For confirmation that the hand that wrote the notes on this folio was Toscanelli's, see Jervis, *Cometary Theory* (see note 71 in Chap. 6), p. 49.

maker of the alleged Toscanelli map,[73] rather than the Toscanelli map and letter that influenced Martellus, Columbus, and Behaim.

In fact, there is evidence that tends to corroborate this conclusion that Behaim was influenced by Martellus rather than Toscanelli and also that the forger of the map attributed to Toscanelli was influenced by Martellus, rather than the other way around.

First with regard to Behaim, one such piece of evidence was alluded to earlier, namely, that there are texts about sea monsters on Martellus map that are derived from the *Hortus Sanitatis* and images of sea monsters on Behaim's globe that are derived from the same source. Toscanelli's alleged map was supposed to have been created in 1474, long before the publication of the *Hortus Sanitatis* in 1491, and it is very unlikely that Martellus and Behaim would decide independently to use the same source for their sea monsters. That is, it seems very likely that Martellus influenced Behaim in this regard. Also, a legend on Behaim's globe just off the eastern coast of Japan is relevant to the question of influence. It begins: *diße jnßel zipangu ligt in orient der welt*, that is, "This island Zipangu lies in the east of the world," supplies various details about the island, and ends with the indication that the information comes from Book 3 of Marco Polo.[74] Yet Behaim's phrasing at the beginning of the legend is somewhat different than Polo's, who writes *Çipangu insula est ad orientem in alto mari distans a littore mangy per milliaria mille quingenta...*, that is, "The island of Japan is in the East in the deep sea, 1500 miles distant from the shore of Mangi." Specifically, Behaim gives a bit more emphasis to the eastern location of the island ("in the east of the world"), and it seems possible that this extra emphasis came from his use of a large world map by Martellus similar to the Yale Martellus map, on which Japan is at the very eastern edge of the

[73] For reconstructions of the Toscanelli map, see Gustavo Uzielli, *La vita e i tempi di Paolo dal Pozzo Toscanelli* (Rome: Ministero della Pubblica Istruzione, 1894), plate 10 (a foldout at the very end of the book); Vignaud, *Toscanelli and Columbus* (see note 37 in Chap. 6), Appendix J, pp. 345–348; and John G. Bartholomew, *A Literary and Historical Atlas of America* (London: J. M. Dent & Sons, Ltd.; and New York: E. P. Dutton & Co., 1911), p. 1. The influence of Martellus on Vignaud's and Bartholomew's reconstructions is obvious but unacknowledged. Sebastiano Crinò, taking a hint from Uzielli, tried to show that the so-called Genoese world map of 1457 was to be identified with Toscanelli's chart: see his "La scoperta della carta originale di Paolo dal Pozzo Toscanelli, che servì di guida a Cristoforo Colombo per il viaggio verso il Nuovo Mondo," *L'Universo: Rivista mensile dell'Istituto Gegrafico Militare* 22.6 (1941), pp. 379–405, and the same author's *Come fu scoperta l'America: a proposito della identificazione della carta originale di Paolo dal Pozzo Toscanelli, la cui copia servì di guida a Cristoforo Colombo per il viaggio verso il nuovo mundo* (Milan: U. Hoepli, 1943). But this suggestion has not been well received. Crinò's latter work is reviewed by William Jerome Wilson in *Geographical Review* 32.1 (1942), pp. 175–177, and for additional bibliography on the reception of his suggestion, see Gaetano Ferro, *La tradizione cartografica genovese e Cristoforo Colombo* (Rome: Istituto Poligrafico e Zecca dello Stato, 1992), pp. 186–187.

[74] See Ravenstein, *Martin Behaim, His Life and His Globe* (see note 72 in Chap. 1), p. 89. Also see Ravenstein, p. 63, who notes that Behaim's other citations of Polo show that he was using Francesco Pipino's Latin version, in some cases that published c. 1484, and in others evidently a manuscript of the version later published by Ramusio in vol. 3 of his *Navigazione* (1556), rather than the German translation published in 1477: Marco Polo, *Hie hebt sich an das puch des edeln Ritters uñ landtfarers Marcho Polo, in dem er schreibt die grossen wunderlichen ding dieser welt* (Nuremberg: Friedrich Creussner, 1477).

image of the world. At the same time, Behaim's phrase decreases the likelihood that he was familiar with Toscanelli's letter, for Toscanelli writes[75]:

> *et non miremini si voco occidentales partes vbi sunt aromata cum communiter dicantur orientales quia navigantibus ad occidentem semper ille partes inueniuntur per subterraneas navigaciones.*
>
> *And do not wonder if I call those places where the spices are western, whereas they are commonly called eastern: because to those that sail by subterranean navigation those places are ever found in the west.*

Behaim's slightly increased emphasis on the eastern location of Japan with respect to the phrasing in Marco Polo is the opposite of what we would expect from someone familiar with the letter attributed to Toscanelli, but accords well with Martellus's image of the world.

There is also evidence that makes it difficult to believe that Martellus was influenced by Toscanelli. The Yale Martellus map covers 280° of longitude, and much of the 80° of longitude that Martellus omits is precisely the area covered by the alleged Toscanelli map, namely, the ocean west of the Cape Verde and Azores and east of Japan: Martellus omits the region that was allegedly the focus of Toscanelli's attention. In addition, neither the legends on Japan in the map of that island in the Florence manuscript of Martellus's *Insularium* (transcribed and translated above) nor the legible parts of the texts on Japan on the Yale map show any sign of influence from the description of the island in Toscanelli's letter. And finally, Martellus did not include the island of Antilia,[76] to which the Toscanelli letter ascribes considerable importance, in any of the surviving manuscripts of his *Insularium*.

Given that the letters and map attributed to Toscanelli are forgeries, and the similarity of their geography to that of the Yale Martellus map, we may confidently conclude that it was the cartographer of the forged map, and also Behaim, who drew from Martellus, rather than the other way around. This conclusion corroborates the great originality and remarkable influence of Martellus's large world map: it shaped the geographical ideas of Columbus that helped inspire his great First Voyage of discovery, guided Behaim in the making of the earliest surviving terrestrial globe, and served as a model for Waldseemüller in the making of his 1507 world map, which gave the name America to the New World.

[75]Vignaud, *Toscanelli and Columbus* (see note 37 in Chap. 6), pp. 296 (Latin) and 280 (English).

[76]For discussion of the cartographic history of Antilia, see William H. Babcock, "Antillia and the Antilles," *Geographical Review* 9.2 (1920), pp. 109–124, and Benjamin Olshin, "The Question of Antilia," in his "A Sea Discovered: Pre-Columbian Conceptions and Depictions of the Atlantic Ocean," Ph.D. Dissertation, University of Toronto, 1994, pp. 189–229.

Conclusions

The 2014 multispectral images of the Yale Martellus map have transformed what was an essentially unstudiable object into one that can be studied in all of its aspects. In the preceding pages, I have explored some aspects of the map, but I hope that the multispectral images will be the foundation of many additional studies of a map that is a profoundly rewarding subject of investigation.

The Yale Martellus map, or rather this group of maps by Martellus of which the Yale map is the sole surviving exemplar, was among the most important of the fifteenth century. Other cartographers before Martellus had incorporated data from Marco Polo into their maps, but Martellus's formation of a cartographic image of eastern Asia based on his study, deductions, and extrapolations from Polo's text seems to be original. Specifically, his integration of Polo's information into Ptolemy's system of latitude and longitude represents a bold expansion by some 100° of longitude—more than a quarter of the earth's circumference—of the area included within that system. Martellus's use of a lost, more complete version of the *Egyptus Novelo* map in his depiction of southern Africa makes his map a unique and precious record of the geography and cartography of southern Africa based on indigenous geographical data and conceptions and is of interest not only as an example of an otherwise almost entirely lost tradition of African mapping but also of the integration of non-European mapping into early modern European cartography.

We have also seen good evidence for the existence of a printed version of Martellus's large world map, a version of the map with a similar high level of detail (and thus inevitably a similar large size) and covering 280° of longitude so as to include Japan, but with several differences in the geographical details, including the arrangement of the islands in the southeastern Indian Ocean. It was this printed version that enabled the diffusion of Martellus's vision of the world and its influence at the highest levels of European culture, on Columbus, Behaim, and Waldseemüller.

C. Van Duzer, *Henricus Martellus's World Map at Yale (c. 1491)*, Historical Geography and Geosciences,
https://doi.org/10.1007/978-3-319-76840-3

Appendix: Equipment and Techniques Used in the Multispectral Imaging of the Yale Martellus Map

Roger Easton, Gregory Heyworth, and Kenneth Boydston

Given the importance to this study of the multispectral images of the Yale Martellus map, an account of how those images were made is necessary.

In July of 2013, a team consisting of Gregory Heyworth (Lazarus Project, then University of Mississippi, now University of Rochester), Michael Phelps (Early Manuscripts Electronic Library), Roger Easton (Rochester Institute of Technology), Kenneth Boydston (CEO of MegaVision), and Chet Van Duzer (Principal Investigator) applied for a grant under the Humanities Collections and Reference Resources Program of the US National Endowment for the Humanities to fund multispectral imaging of the Yale Martellus map. The grant application was successful, and the team visited the Beinecke Rare Book and Manuscript Library in New Haven for 10 days in August of 2014 to make the images.

Multispectral imaging is a powerful tool for recovering text in manuscripts, printed books, and maps that have been obscured by fading, water damage, over-painting, palimpsesting, and cockling (wrinkling). The use of the nomenclature "multispectral imaging" and "hyperspectral imaging" refers to different numbers and widths of the spectral bands of the imagery. Unfortunately the namings are not standardized, so it is more convenient to refer to the collection and processing as "spectral imaging." The process is based upon the observation that pigments reflect, absorb, and occasionally emit different colors of light in different amounts. This is also true for invisible wavelengths of light: different pigments respond differently to ultraviolet light, which is of a wavelength shorter than the eye can see, and to infrared light, which is of a wavelength longer than the eye can see. By examining images collected under different colors of light, it is possible to combine these images to make damaged texts visible.

Spectral imaging has been in use for many years; it has been used in astronomy since the 1950s to classify stars based on the relative emission in blue, green, and ultraviolet bands. However, the technologies for illumination and imaging have only recently advanced to the point where it has become possible to collect spectral images with sufficient spatial and spectral resolution that it may be usefully applied to historical manuscripts and other documents. A wide range of light-emitting diodes (LEDs) is now available to illuminate the object at exactly the desired wavelengths, lenses have been designed and fabricated that can create images over the wider range of wavelengths emitted by the LEDs, and 50 megapixel sensors are available for converting the image into computer data that may be processed and studied.

The large size, degraded condition, and iconic status of the Yale Martellus map necessitated considerable planning to ensure that a single imaging session would be successful. Planning began over a year before the imaging session commenced. All of the equipment required for imaging had to be shipped from various locations to the Beinecke Library. It would be impossible to test the entire system until it was assembled at the library. A

MegaVision multispectral imaging system was supplied by the Lazarus Project, additional equipment and software to facilitate imaging an object the size of the Martellus map were supplied by MegaVision, and various accoutrements were supplied by different members of the imaging team.

Once Chet Van Duzer had secured funding for the project, the imaging team congregated at the Beinecke Rare Book and Manuscript Library at Yale University in August of 2014 to image the map. To support the map for imaging, the team set up a large easel supplied by MegaVision that was inclined at a 10° angle so that gravity would keep the map in place. The map's large size (approximately 2 meters wide by 1.25 meters tall) entailed that it had to be imaged in overlapping sections called "tiles," 11 tiles across by 5 down. The need to precisely maintain the positions of the LED lamps relative to the camera entailed that the map itself, rather than the lights and camera, had to be moved in order to bring the different tiles into the camera's field of view. The tray of the easel was moveable, so that the map could be translated in the "*x*" (left to right) and "*y*" directions (top to bottom). The increments of motion were marked on the easel so that manually moving the easel to the various image tile locations could be performed repeatably and easily. To eliminate risk and facilitate ease of motion, the easel included a counterweight to the weight of the map, so that vertical movement of the map was effortless.

It is important to ensure that the distance from the map surface to the camera does not change, despite irregularities in the map's surface and in the canvas frame supporting it. For this reason, the camera was mounted on a controllable device that allowed it to be moved toward or away from the map (along the "*z*" axis) in small increments. This device is a commercially available "stack shot" designed to combine images taken at different distances with the same focus setting to construct images with larger apparent depth of field. After focusing the camera on a planar reference surface (in our case, a machined sheet of Spectralon®, a very white material) placed at the nominal location of the map surface, the distance to the map is set by setting two sets of parallel lines of laser light, one pair oriented horizontally and one vertically, to coincide at the selected distance. After the map is translated to a new tile, the stack shot is used to move the camera toward or away from the map until the pairs of lines again coincide. This step is quickly implemented and ensures that the image of each tile is focused and has the same magnification and therefore may be digitally "stitched" to make images of the entire map.

The map was illuminated using LEDs at 12 spectral bands centered at wavelengths from the ultraviolet part of the spectrum at 365 nm through the visible range and to the infrared part at 940 nm. Unlike incandescent lamps, which generate light as a by-product of resistive heating, light from LEDs is the outcome of an electronic process. By consequence, very little heat is produced by LEDs, so that they are much safer for use in imaging of manuscripts. In addition, LEDs emit light over narrow bands of wavelengths, with bandwidths of the order of 20–40 nm, which makes them very useful for spectral imaging. Incandescent lamps emit light over virtually the entire range of the spectrum. To obtain the narrowband illumination necessary for spectral imaging, the broadband illumination from incandescent lamps must be filtered to pass only the desired range of wavelengths, which makes the image collection process much more complicated.

In addition to the images of reflected light, previous experience has demonstrated the value of images of light generated by fluorescence within the object for recovering faded or erased writings. Fluorescence is the longer-wavelength light produced within a material after having absorbed shorter-wavelength light. The most common example is the visible emission by a material after absorbing ultraviolet light. Broadband images of ultraviolet fluorescence were important in the image processing of the *Archimedes Palimpsest*, as the contrast of the erased text was significantly better at these wavelengths.[1] The erased text on one of the manuscripts in the palimpsest was recoverable only from combinations of fluorescence images collected through red, green, and blue filters. This provided the impetus to collect spectrally filtered fluorescence images. To automate capture of images which discriminate between reflected excitation illuminant and emitted fluorescence, the camera has a pair of software-controlled filter wheels to carry filters that selectively pass UV, broadband visible, red, orange, green, and blue light to the camera when the object is illuminated by ultraviolet or blue light.

The images of the object in the various wave bands are collected by a lens that was specially designed for spectral imaging by J. Brian Caldwell of Caldwell Photographic, Inc. The lens has a focal length of 120 mm and maximum aperture of f/4 and is made not of ordinary lens glass but of fluorite and fused quartz, which allow it to produce sharply focused images at the same focal plane from the ultraviolet end of the spectrum through the visible and infrared bands. Lenses made of ordinary lens glass cannot focus infrared light on the same plane as visible light and do not transmit ultraviolet light well—and thus are ill suited to spectral imaging.

The unavoidable diffraction of the light sets the ultimate limit on the resolution of the imaging system. It also ensures that only a limited depth of field appears to be in focus in the recorded images. A larger range of distances from the lens appear in focus if the diameter of the lens aperture is reduced; the common term is that the lens is "stopped down." To record

[1] See William A. Christens-Barry, Roger L. Easton, Jr., and Keith T. Knox, "Imaging and Image-Processing Techniques," in Reviel Netz et al., eds., *The Archimedes Palimpsest* (Cambridge and New York: Published for the Walters Art Museum by Cambridge University Press, 2011), vol. 1, pp. 182–215; and Roger L. Easton, Jr., William A. Christens-Barry, Keith T. Knox, "Ten Years of Lessons from Imaging of the Archimedes Palimpsest," *Commentationes Humanarum Litterarum* 129 (2011), pp. 5–34.

the same amount of light, the exposure time must be longer for smaller apertures. The same lens aperture was used for all reflected light captures so that the image quality was approximately equal, but the different strengths of the LEDs at the different wavelengths required that the exposure time and/or light power had to be changed to ensure similar exposures. For the reflectance images, the lens was stopped down to f/12.5 to maximize the depth of field and the exposure times ranged from 0.6 s to 11.9 s. For the much fainter fluorescence images, the lens was set at f/5.6, and most of the exposure times were 5, 10, or 20 s, though the exposure through the blue filter for blue illumination was only one-half second.

The light in the images captured by the MegaVision camera is recorded by a KAF-50100 charge-coupled device array from TrueSense Imaging (now ON Semiconductor) with 8176 × 6132 pixels at the image plane. Each pixel element is 6 micrometers square, so the complete array covers 49.1 mm × 36.8 mm at the image plane. The sensor converts light photons to electrons, which are counted to provide a measure of the illumination at each point in the image field. The sensor can record as many as 40,000 photoelectrons at each pixel, but this number is converted to a "digital count" or "gray value" by quantization to 14 bits, for a total of $2^{14} = 16{,}384$ possible levels, ranging from "0" for no exposure ("black") to "16,383" for the maximum ("white"). Given the dynamic range (signal-to-noise ratio) of the sensor, 12 bits are adequate to characterize the signal, so the 14 bit signal is appropriately mapped to 12 bits as a part of image capture.

Other than image calibration, correction, and construction of accurate color renditions of the image tiles, most of the image processing was performed in the *ENVI* software package from Exelis Visual Information Solutions. This package was designed for environmental remote sensing, but includes a set of algorithms that are useful for imaging of historical artifacts. After all, imaging the map is actually a remote sensing problem, just with a much smaller distance between the object and the camera. In addition to managing image capture, MegaVision's PhotoShoot software performed preliminary image calibration and correction and generated very accurate color renditions from the reflectance images captured under the seven visible wave bands.

After the set of images for all 55 tiles have been collected, the images must be calibrated, analyzed, and processed to enhance the visibility of the features of interest. The calibration step compares the gray values of each pixel to a reference "white" provided by a piece of Spectralon® from Labsphere, which accurately reflects 98% of the incident light over a very wide range of wavelengths. The calibration converts each integer gray value to a floating-point number between "0" and "1," corresponding to the approximate reflectance at each pixel. The same calibration is used for the fluorescence images, though the emission may result in calibrated values larger than "1."

The calibrated images were then combined to form an image "cube" (called a "stack" in some quarters), which is a three-dimensional array of data with coordinates $[x,y,\lambda]$, where x and y are the spatial coordinates and λ is the central wavelength of each imaging band. In this way, each horizontal "slice" of the $[x,y]$ plane at a specific wavelength is the specific image band with that wavelength. The various slices through the cube are processed and combined using various algorithms, many of which developed during the *Archimedes Palimpsest* project. Some of the methods are used to "precondition" the image data, such as to attenuate noise or to reduce the dynamic range of the data in regions with different background colors, such as water and land.

On very rare occasions, the feature of interest might be sufficiently visible in a single spectral image band. Because that happy condition is so rare, it is generally necessary to "process" the images to enhance the visibility of the feature. The concept of image processing may be distilled down to the concept of finding a combination of the image bands such that the feature of interest becomes visible. Image processing algorithms fall roughly into two categories: "deterministic" and "statistical."[2] The deterministic combinations of image bands were determined a priori and do not depend on the conditions in a particular image. The most common deterministic image processing is formation of a "false-color" (or "pseudocolor") image. Instead of putting the image bands collected under red, green, and blue illumination into the red, green, and blue bands of a color image, different bands are selected. For example, it is occasionally useful to put the images of ultraviolet fluorescence through the red, green, and blue filters into the red, green, and blue channels. Such a combination often enhances the visibility of faded text in a palimpsest.

The calculations of statistical image processing are based on the specific image cube, determined largely by the number of pixels with each gray value—the so-called histogram of the image. In a single band, the histogram is a graph of the number of pixels in the image with each of the 4096 gray values. The gray value that occurs most frequently in the image will have the largest value in the histogram. In a two-band image, the histogram is a graphical plot of the number of pixels with each pair of gray values. For example, the number of pixels that have gray value of 100 in the first image and value of 120 in the second is placed in the bin indexed by [100,120] in the two-dimensional histogram. For the Martellus images, the statistical processing often used the 12 reflective bands plus 10 fluorescence bands, so the histogram occupies a 22-dimensional space. Though impossible to visualize, a space with so many dimensions may be evaluated in a computer quite easily, though the process often is quite time consuming.

[2] Roger L. Easton, Jr. and David Kelbe, "Statistical Processing of Spectral Imagery to Recover Writings from Erased or Damaged Manuscripts," *Manuscript Cultures* 7 (2014), pp. 35–46.

The statistical processing algorithm used most often and with the greatest success in the Martellus project was "principal component analysis" (PCA), which constructs equivalent set of N image bands from an N-dimensional image cube. The first of the new set of bands is defined by the axis in the N-dimensional space that spans the widest possible dynamic range of the image data. Once the axis is specified, then all image pixels are "projected" onto this axis, forming a one-dimensional histogram of "gray values." One end of the histogram is specified as "white" and the other as "black," so that the position on this axis of a projected pixel determines its gray value in the new image of this principal component. The axis in the N-dimensional space that is orthogonal to the first axis that spans the widest possible dynamic range in that hyperplane defines the second principal component. The image pixels in the N-band histogram are again projected onto this new axis, and the position relative to the extreme values on that axis determines the gray value of that pixel in that principal component band. This process of finding each subsequent axis as orthogonal to the previously determined axes, followed by projecting the pixels in the N-band histogram to determine the gray value in the principal component, is continued until N equivalent images are determined from the original N-band data.

The dynamic range of the data decreases with increasing order of the principal components, which means that the high-order principal components that are evaluated last tend to span a very small dynamic range of gray value. For this reason, PCA often is used in imaging situations to discard the random noise in an image that appears in the high-order components. In the calculations for the Martellus map, the high-order principal components sometimes contain very subtle features of the map (such as faded text), rather than just random noise. In fact, one of the useful PCA bands of the Martellus map was band #22 of a 22-band image cube. This outcome is unique in the experience of this imaging team and demonstrates that the contrast of these features relative to the backgrounds is smaller than the random variations in the map reflectance or the random variations generated during imaging. This suggests that these features could not be segmented from a small number of spectral bands.

Though the principal components are orthogonal, it is again a rare happy occurrence when the feature of interest appears in a single principal component. More often, the different image features are more visible in a few different principal components. In such a case, we have found it useful to construct a pseudocolor image from three different principal component bands. Yet another step is often effective; the color image is manipulated in Adobe Photoshop® or similar software package. The color image may be specified as red, green, and blue Cartesian coordinates (RGB) or as "lightness," "saturation," and "hue" in a cylindrical coordinate system. Rotation of the hue angle of the pseudocolor image often enhances the visibility of the feature of interest.

We noticed that in many places the background pattern "noise" of the blue filtered light image and the background pattern noise of the orange filtered UV light image were in some sense complementary. We thought perhaps there might be a way to blend the two images so that the corresponding background patterns might cancel, while the text common to both would mutually reinforce; this was due to the speculation that the background pattern was due to the texture of the map surface.

After running a median-type filter to minimize random noise, we normalized the two images by adjusting the tone curves (keeping them linear) so that the image histograms were roughly equal and the average of the histogram was about mid-tone gray. We then took the "inverse" of one image (perhaps better understood as a "negative image") and combined the images as layers in Photoshop. Using the layers blending slider, we set the blending to visually minimize the background "noise." For reasons not entirely understood, the resulting blended image revealed a fair amount of the same text that was revealed by principal components, though with considerably less contrast, and surprisingly, a small but appreciable amount of additional text not visible in the principal component images was revealed.

Much of the image processing was performed on the individual captured image tiles. In order to appreciate the map as the single entity that it is, the image tiles needed to be stitched together. The color images derived from the visible multispectral data were stitched using PTGui software. After stitching some of the principal component images, it was observed that tile boundaries were more visible than was desired, so aesthetic adjustments to the rendered image tiles were made to minimize the boundary differences.

Since three different processes revealed three different sets of texts, a final step was undertaken to merge the results of the processes. There appeared to be no obvious automated method, so the three stitched images were added as layers to the stitched color image and Photoshop's layer mask tool was used to combine revealed texts from each of the layers in a single image.

Index of Names

A

Aalto, Pentti, 92
Akerman, James R., 5, 10
Albertin de Virga's world map, 32, 36, 152, 153
Alexander, Jonathan J. G., 18, 19
Alexander, the King, 95, 96
Alfonso V of Aragon, 158, 161, 163
Almagià, Roberto, 4, 9, 11–13, 15, 23, 26, 30, 32, 35, 53, 82, 83, 115, 176
Anastos, Milton Vasil, 24, 53
Angeli da Scarperia, Iacopo, 4
Anliker, Ernst, 3
Apian, Peter, 80, 82
Arabia Deserta, 121
Aranges montes, 114
Archimedes Palimpsest, 197, 199
Aristotle, 71, 87, 186
Arnold the Saxon's *De floribus rerum naturalium*, 64
Attavante degli Attavanti, 19
Aulesa, Thomas, 159
Auty, Robert, 3
Axelson, Eric, 12, 23, 134

B

Babcock, William H., 191
Babicz, Józef, 3, 4, 30, 161
Bacon, Roger, 186
Bangala Provi, 100
Banks, S. E., 15
Barney, Stephen A., 41, 115
Beach, J. A., 41, 115
Beans, George H., 26
Becaria, Francesco, 162, 163
Beckingham, Charles F., 149, 157, 185
Bedini, Silvio A., 3
Behaim, Martin, 27, 28, 30, 39, 41–42, 80, 85, 86, 167, 170, 175, 190, 191, 193
Behaim's globe, 30, 34, 77, 79, 82, 125–130, 132, 168–174
Bemoim, John, 115, 116
Berggren, J. L., 34
Berghof, Oliver, 41, 115
Berlinghieri's world map, 31
Binns, J. W., 15
Blackburn, Mark, 160
Boese, H., 41, 78, 87
Bonacker, Wilhelm, 3
Bond, William H., 12
Böninger, Lorenz, 3, 5
Bordone's, Benedetto map, 82
Borso d'Este, 161, 163
Bostock, John, 60
Boydston, Kenneth, 195
Bracke, Wouter, 9
Braukämper, Ulrich, 144–146
Brown, Clifford M., 161
Bumi Jeléen, 115
Buondelmonti, Cristoforo, 9, 10, 12
Buondelmonti's *isolario*, 8
Buringh, Eltjo, 8, 35
Buron, Edmond, 179, 186
Burri, Renate, 18

C

Cabot's map, 82
Caldwell, J. Brian, 197
Campbell, Tony, 22, 23, 27, 28
Cão, Diogo, 17, 130–133, 184, 185
Casas, Bartolomé Las, 179, 180, 182, 183
Cavallo, Guglielmo, 5, 15, 149, 151, 161
Caverio, Nicolo de, 36, 45, 128, 130
Caverio's map, 36, 45, 46, 119, 125–127, 129, 131, 133, 134
Ceccarelli, Paola, 9
Celer, Quintus Metellus, 61
Chinchitila Provintia, 95
Christens-Barry, William A., 197
Cipangu do wächßt vil gold, 86
Cisne, John L., 35
Civai, Allesandra, 3
Clark, Willene B., 41
Colón, Fernando, 175, 182, 183, 186
Columbus Chart, 152
Columbus, Christopher, 18, 23, 24, 28, 34, 53, 61, 79, 80, 175, 176, 178–191, 193
Contarini, Giovanni Matteo, 80, 81
Contarini's map, 31, 98, 100, 145, 167
Cordeiro, Luciano, 17
Crawford, O. G. S., 140, 142, 145
Crone, G. R., 126, 174

C. Van Duzer, *Henricus Martellus's World Map at Yale (c. 1491)*, Historical Geography and Geosciences,
https://doi.org/10.1007/978-3-319-76840-3

D
d'Ailly, Pierre, 179, 186
d'Avezac, M., 42
Davidson, Miles H., 181–183, 185, 186, 189
De insulis et earum proprietatibus, 9, 63
de la Mare, Albinia, 18, 159
de Meyïer, K. A., 13
Dias, Bartolomeu, 12, 17, 18, 23, 36, 39, 130–135, 179, 184, 185
Dines, Ilya, 18, 41
Dörflinger, Johannes, 3
Dotti, G., 19
Dümmler, Ernst, 47

E
Eames, Wilberforce, 28, 119
Easton, Roger, 195, 197, 199
Ebstorf *mappamundi*, 52, 162
Edson, Evelyn, 4, 12, 22, 179
Egyptus novelo map, 112, 117, 136–165, 174, 193
Enenkel, Karl A. E., 42
Engelmann, W., 30

F
Falchetta, Piero, 32, 38, 96, 145, 151, 155, 157, 174
Falconer, William, 25
Faldini, Luisa, 54, 58
Faye, Christopher U., 12
Ferro, Gaetano, 54, 58, 190
Fischer, Joseph, 3–6, 17, 48
Fischer, Theobald, 38
Fra Mauro, Fra, 80, 96, 157, 163, 174
Fra Mauro's *mappamundi*, 32, 82, 145, 151, 155–157, 175
Franklin, Michael W., 19
Friedman, John B., 90

G
Garretson, Peter P., 164
Gentile, Sebastiano, 4–6, 8, 10, 26, 32, 82, 160
Gerini, G. E., 70, 72
Germanus, Donnus Nicolaus, 3, 5, 6, 17, 30, 32, 53, 57, 161
Gervase of Tilbury's *Otia imperialia*, 15
Gilbert, Humphrey, 61
Gioia, Professor, 26
Giovanni de Maestro Antonio, 160
Goldstein, Thomas, 137, 188
Gow, Andrew, 96
Graham, Victor, 116
Great Khan, 90, 94, 98, 101, 185
Greenia, George D., 161
Grierson, Philip, 160
Guido da Pisa, 64

H
Haebler, Konrad, 26
Hakluyt, Richard, 61
Hallberg, Ivar, 76, 90, 92, 96, 99
Hamilton, Bernard, 149
Hamilton, Hans Claude, 25
Hammer, Heinrich, 3
Harris, Elizabeth, 113, 114
Heawood, Edward, 31
Henry the Navigator, 62
Herbermann, Charles George, 48
Hereford *mappamundi*, 52–53
Hessler, John W., 32, 33, 48, 54, 57, 58
Heyworth, Gregory, 195
Hind, Arthur M., 26
Hirsch, Bertrand, 140, 142–146, 155, 157
Historia de las Indias, 182–183, 188
Historia rerum ubique gestarum, 8, 13, 15, 16, 22, 55, 62, 182
Holmes, John, 15
Holzschuher, Georg, 27
Hortus Sanitatis, 40, 41, 64–67, 71, 78, 88, 112, 113–116, 168, 169, 171, 174, 190
Hülsen, Christian, 26

I
Insularium illustratum Henrici Martelli Germani, 13
Isidore's *De natura rerum*, 48–53
Isidore's *Etymologiae*, 40, 53, 64, 90, 92, 114
Isidore of Seville, 80, 92
Ixola de cimpagu, 32

J
Jerome, St., 19
Jervis, Jane L., 181, 189
João II of Portugal, 115, 134
Johnston, Mark D., 178
Jones, A., 34
Jorath, 64

K
Kalkwiek, K. A., 13
Karrow, Robert W., 4
Kelbe, David, 199
Kelley, James E., 180
Kimble, George H. T., 38
Kish, George, 9, 32, 82, 83, 175
Klebs, Arnold C., 40
Konrad von Megenberg, 78
Kristeller, Paul Oskar, 12

L
Lach, Donald F., 110
Lambert of St. Omer's *Liber floridus*, 52
Landry, Judith, 18
Lang, A. W., 27, 28
Leardo's *mappaemundi*, 149–151
Lewis, W. J., 41, 115
Liber insularum Archipelagi, 9, 10
Listri, Massimo, 18
Lops, R. L. H., 116
Lowe, Katherine J. P., 18

M
Madden, Frederic, 15
Magnaghi, Alberto, 179
Major, R. H., 176
Mannoni, Laura, 140
Manuel, King, 18
Markham, Clements R., 182
Marques, Alfredo Pinheiro, 3
Martellus, Henricus, 2–42, 119, 126, 137, 139, 143, 151, 154, 158, 163, 165, 193
Martellus-Rosselli map, 22–24, 30, 31, 34–36, 38, 40, 47, 55, 74–76, 87, 88, 90, 92–94, 98–100, 110, 111, 115, 127, 130–134, 150, 167, 168, 172
Martellus's *Insularium illustratum*, 4, 6, 9–18, 22
Martins, Fernão, 181–183, 188
McCrindle, J. W., 67, 105
Messerli, Luisa Rubini, 3, 4
Meurgey de Tupigny, 13, 14
Milanesi, Gaetano, 19
Milanesi, Marica, 9, 54, 58
Morison, Samuel E., 188
Moule, A. C., 99, 101, 109
Müllenhoff, Karl, 60
Münster, Sebastian, 80

N
Narducci, Enrico, 5
Neckam, Alexander *De naturis rerum*, 71
Nepos, Cornelius, 60
Nicander's *Theriaca*, 114
Nicholas of Cusa, 6
Nordenskiöld, A. E., 96, 149
Nunn, George E., 176, 180

O
Obbema, P. F. J., 13
Olshin, Benjamin, 191
Ongania, F., 38, 151

P
Paviot, Jacques, 150
Payne, Joseph Frank, 40, 41
Pekkanen, Tuomo, 92
Pelliot, Paul, 63, 76, 88, 89, 98, 101, 108
Penedo das Fontes, 12
Petz, Johann, 27
Phelps, Michael, 195
Phillips, William D., 178
Piccolomini, Enea Silvio, 8, 13, 15, 16, 22, 55, 61, 62
Pico della Mirandola, Gianfrancesco, 42
Pietro del Massaio, 6, 140, 142, 143, 160, 161
Pinzón, Martín Alonso, 178
Pipino, Francesco, 35, 87, 99, 190
Pletho, 24, 53
Pliny, 41, 80, 90, 92, 116, 186
Pliny's *Naturalis historia*, 59, 60, 87, 186
Polo, Marco, 12, 32, 34, 35, 63, 67–69, 72, 74–76, 79, 80, 82–92, 94–99, 100–103, 106–111, 172, 174, 175, 184, 190, 191, 193
Pomponius Mela, 61, 80
Prester John, 146, 147, 149–152, 154, 184, 185
Ptolemy's *Geography*, 4–8, 13, 15, 22, 28, 29, 31, 34, 42, 47, 57, 62, 67, 69, 72, 93, 104, 106, 117, 119, 137, 140, 143, 158–161, 163, 164, 187
Ptolemy's world map, 12

R
Raff, Thomas, 47
Ragone, Giuseppe, 9
Randles, W. G. L., 12, 17, 140, 143–146, 178–180
Ranzano, Pietro, 142
Ravenstein, E. G., 17, 27, 80, 85, 86, 126–133, 170–172, 174, 175, 184, 185, 190
Reichert, Folker, 82
René II, 41, 42
Ribes, Jacme, 162, 163
Richard, J., 150
Richardson, Carol M., 19
Riley, H. T., 60
Ringmann, Matthias, 42, 48
Ristow, Walter W., 52
Rombolo, Pietro, 165
Romer, Frank E., 61
Romulo da Messina, Pietro, 142
Rosen, Mark S., 24
Ross, D. J. A., 96
Rosselli, Francescso, 4, 20–23, 26, 28, 80, 137, 139

S
Sanz, Carlos, 28, 119
Schmidt, Charles, 29, 42
Schoeffer, P., 40
Schöner, Johann, 65
Schwarz, E. H. L., 12
Scott, Jeremy, 32
Seneca, 186
Shipley, John, 187
Shirley, Rodney W., 20, 31, 80
Sighinolfi, L., 119
Signorini, Maddalena, 18
Silvestri, Domenico, 9, 63, 77
Skelton, R. A., 24–27, 30, 31, 35, 38, 52, 119, 137, 162, 175, 176, 178–180
Solinus, 80
Solomon, King, 178
Sonetti's *isolario*, 9
Soulsby, Basil H., 3
Spufford, Peter, 160
Stevenson, E. L., 17, 27, 36, 61, 104, 106, 127–134, 151, 154
Stopel, Jacob, 110
Strabo, 24, 25, 53

T
Thebet provincia, 102
Theodulf, 47
Thomas of Cantimpré, 79, 87
Thomas of Cantimpré's *De naturis rerum*, 41, 64
Thomas, St., 109, 110

Toscanelli map, 181, 183, 189–191
Toscanelli, Paolo dal Pozzo, 151, 181–191

U
Uzielli, Gustavo, 181, 189, 190

V
van der Meulen, W. J., 69, 72
Van Duzer, C., 19, 22, 28, 32, 47, 80, 93, 160, 195, 196
Varthema, Ludovico di, 109, 110
Vasco da Gama, 135, 137
Verdenhalven, Fritz, 25
Vespucci, Amerigo, 42, 54, 55
Vignaud, Henry, 181–186, 188, 189, 191
Vincent de Beauvais, 64
Vincent of Beauvais's *Speculum naturale*, 64, 114
Vitelli, Camillo Maria, 5
von Harff, Arnold, 65
von Wieser, Franz, 32, 36, 45, 48, 125, 176

W
Waldseemüller, Martin, 27, 28, 32, 36, 40, 42, 46–48, 53–57, 60–63, 64–66, 74, 80, 88–90, 94, 98–111, 115, 116, 119, 122, 125, 126, 128, 130–132, 191, 193
Waldseemüller's *Carta marina*, 33, 54, 93, 116, 117, 143
Waldseemüller's 1507 map, 17, 32, 34, 42, 45, 53, 54, 56–58, 60, 62, 63, 66–73, 75–79, 84–88, 90, 92, 93, 97, 112–114, 121, 143, 150, 167, 168, 172
Washburn, Wilcomb E., 34
Wawrik, Franz, 3
Weiss, Roberto, 4
Westrem, Scott D., 53, 95, 151
Williamson, J. A., 180
Wolf, Anne Marie, 178
Wolf, Thomas, 42
Woods, Kim W., 19
Woodward, David, 9

Y
Yale Martellus map, 3, 8, 20, 24, 26–31, 34, 36, 38, 39, 41, 44–117, 119, 125, 126, 128, 132, 133, 135, 137–139, 143, 144, 148, 150, 151, 154, 165, 167–172, 174–176, 178–181, 191, 193, 195
Yule, Henry, 34, 63, 68, 75, 79, 83, 85, 87–90, 94–99, 102, 106–110, 175

Z
Zinpanghu insula, 83
Zorzi, Alessandro, 176, 177

Index of tiles of Yale Martellus Map

C01R01, 48
C01R02, 48
C01R04, 48
C02R01, 51
C02R02, 62
C02R05, 49
C03R01, 51
C03R02, 87
C03R03, 87, 101, 112–116, 126, 127, 130, 131, 147
C03R04, 127–131
C03R05, 132, 133
C04R03, 64, 116, 119, 121
C04R04, 64, 65
C04R04b, 65
C04R05, 135, 143
C05R01, 51
C05R02, 59, 87
C05R03, 63, 64
C05R04, 64, 66, 67
C05R05, 49, 68, 135, 143, 147, 148
C06R01, 51
C06R02, 65, 75, 83, 87–92, 105, 107, 109, 126, 128
C06R03, 104
C06R04, 67, 69, 70
C06R05, 68
C07R01, 51
C07R02, 93–95, 103
C07R03, 106
C07R04, 70–72, 74
C08R02, 95, 96, 98, 101–103
C08R03, 102, 106
C08R04, 75, 76, 107–110
C08R04b, 109
C09R01, 50
C09R02, 98–100
C09R04, 77
C09R05, 49
C10R01, 83
C10R02, 79, 80, 83
C10R03, 76, 77
C10R04, 54, 76, 77
C10R05, 54
C11R02, 50
C11R04, 50, 54
C11R05, 54

C. Van Duzer, *Henricus Martellus's World Map at Yale (c. 1491)*, Historical Geography and Geosciences,
https://doi.org/10.1007/978-3-319-76840-3

Index of Manuscripts

B

Bern, Burgerbibliothek, MS 144/2, 18

C

Chantilly, Bibliothèque du Musée Condé, MS 698 (483), 13–15

F

Florence, Biblioteca Medicea Laurenziana, Pluteo 29.25, 10–12, 32, 33, 82
Florence, Biblioteca Nazionale Centrale, Banco Rari 30, 189
Florence, Biblioteca Nazionale Centrale, Magliabechiano XIII 16, 5, 13, 31, 106, 164

G

Ghent, Universiteitsbibliotheek, MS 92, 52

L

Leiden, Leiden University Library, MS VLF 23, 13, 14
Lisbon, Arquivos Nacionais da Torre do Tombo, MS 161/7, 18–19
London, British Library, Add. MS 15760, 15–18
London, British Library, Egerton MS 73, 130
London, British Library, Harley MS 7182, 159–160, 163

M

Madrid, Biblioteca Nacional de España, MS Res. 255, 159–163
Minneapolis, James Ford Bell Library, MS B 1475 fMA, 10, 12–13
Modena, Biblioteca Estense Universitaria, MS Lat. 463 = α.X.1.3, 161
Monistrol de Montserrat (Spain), Abadia de Montserrat MS 1, 154

N

Naples, Biblioteca Nazionale Vittorio Emanuele III, Vind. lat. 50, 35, 98
New York, New York Public Library MS 97, 104, 106

P

Paris, Bibliothèque nationale de France, Rés. Ge B 696, 150
Paris, Bibliothèque nationale de France, MS Espagnol 30, 131
Paris, Bibliothèque nationale de France, MS lat. 4802, 117, 140, 141, 159, 161
Paris, Bibliothèque nationale de France, MS lat. 6556, 139
Paris, Bibliothèque nationale de France, MS lat. 8878, 152
Parma, Biblioteca Palatina, Carta nautica no. 1612, 52
Prague, Knihovna Pražské Metropolitní Kapituyí, G 28, 86

S

Salamanca, Biblioteca General de la Universidad, MS 2586, 160

V

Vatican City, Biblioteca Apostolica Vaticana, Borgia XVI, 152
Vatican City, Biblioteca Apostolica Vaticana, MS Ottob. lat. 1417, 151
Vatican City, Biblioteca Apostolica Vaticana, MS Urb. graec. 82, 47
Vatican City, Biblioteca Apostolica Vaticana, MS Urb. lat. 277, 117, 140
Vatican City, Biblioteca Apostolica Vaticana, MS Vat. lat. 5699, 117, 140, 142, 159, 161
Vatican City, Biblioteca Apostolica Vaticana, Vat. lat. 7289, 4–5, 31
Vicenza, Biblioteca Civica Bertoliana, MS 598 A, 151

C. Van Duzer, *Henricus Martellus's World Map at Yale (c. 1491)*, Historical Geography and Geosciences,
https://doi.org/10.1007/978-3-319-76840-3

MIX
Papier aus verantwortungsvollen Quellen
Paper from responsible sources
FSC® C105338

If you have any concerns about our products,
you can contact us on
ProductSafety@springernature.com

In case Publisher is established outside the EU,
the EU authorized representative is:
Springer Nature Customer Service Center GmbH
Europaplatz 3, 69115 Heidelberg, Germany

Printed by Libri Plureos GmbH
in Hamburg, Germany